This book is printed in large type to aid our visually impaired readers.

Front and back cover paintings: William North
Photo opposite courtesy of *Naples Daily News/The Banner*

ISBN 978-1-935751-06-9
Florida Citrus in Fine Art: Paintings of the Golden Harvest

Published by
Scribbulations LLC
www.scribbulations.com

Printed in the United States of America

Florida Citrus In Fine Art

Paintings Of The Golden Harvest

WILLIAM NORTH

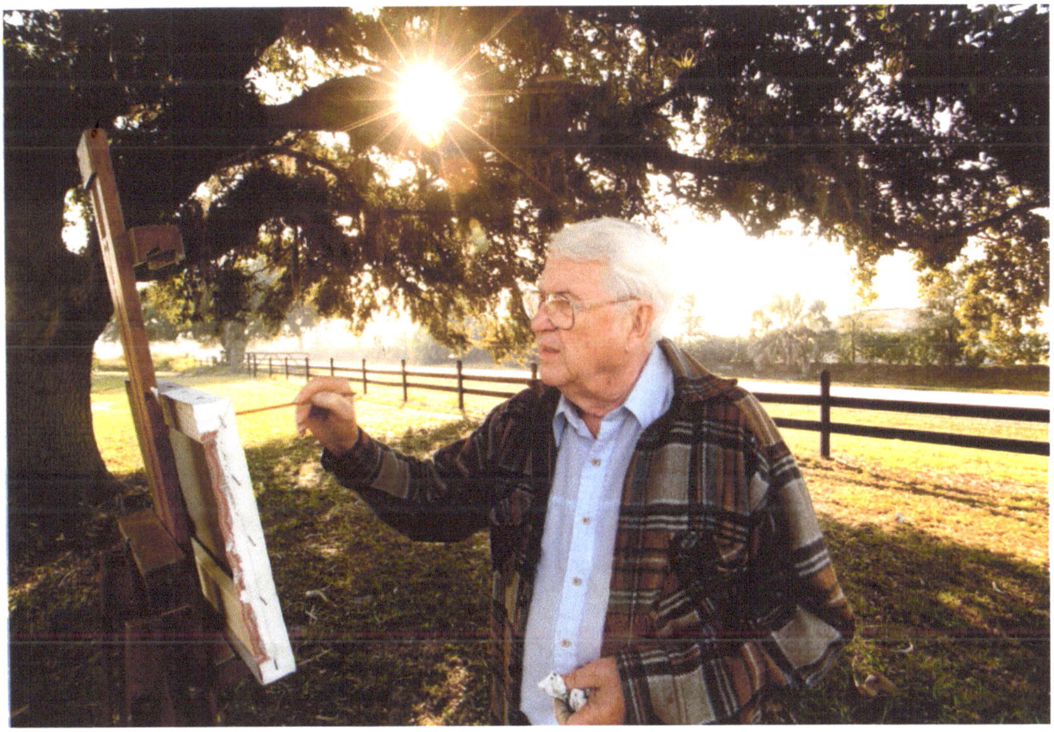

PHOTO COURTESY OF NAPLES DAILY NEWS/THE BANNER

WITH KAREN C. NORTH & COLLEEN R. NORTH

The Family

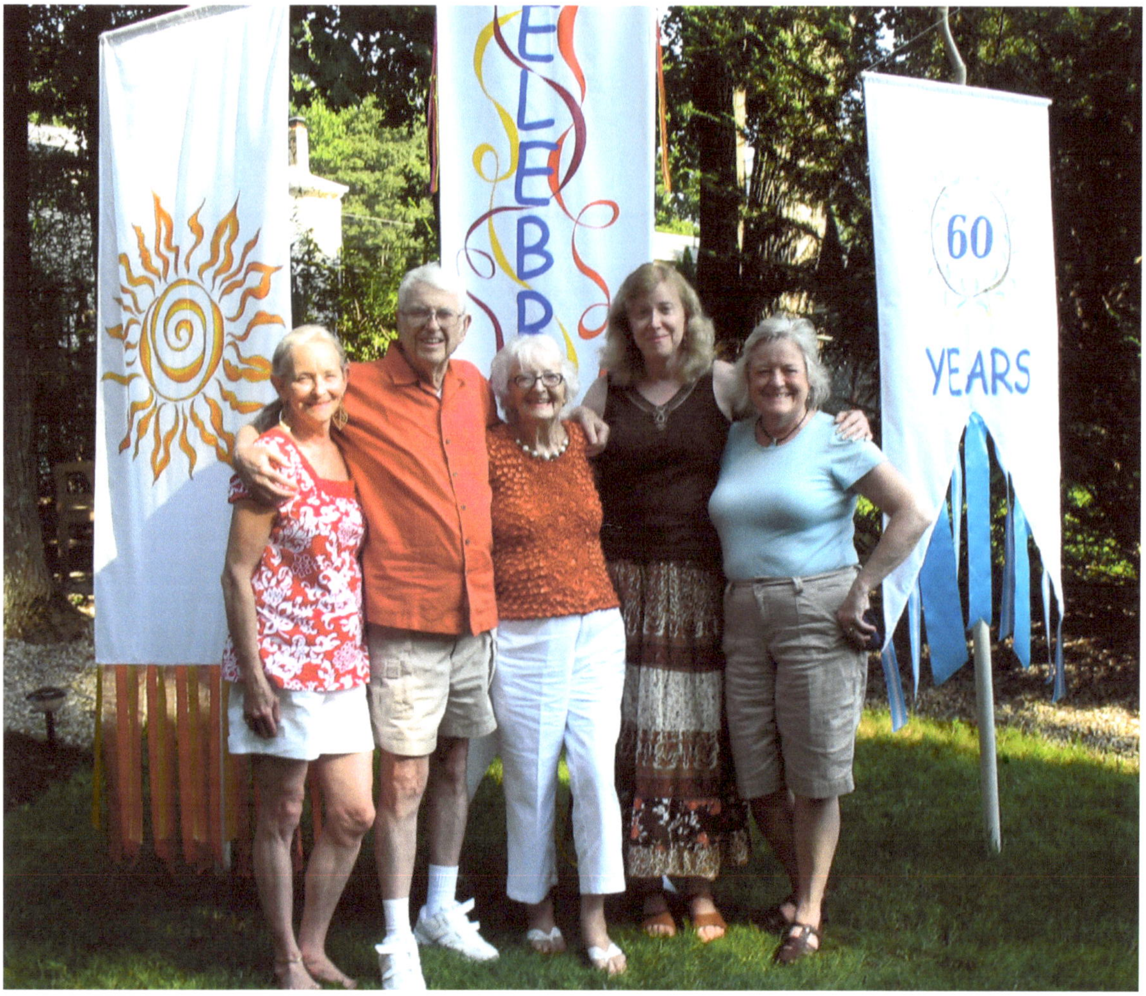

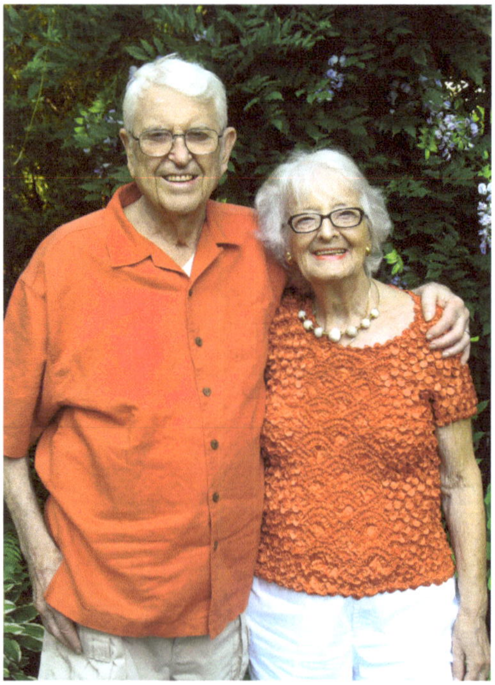

Top Photo from left to right: Debby, Dad, Mom, Karen and Colleen Rebecca.
Left Photo: Bill and Colleen
Bottom Photo: Rich and Debby

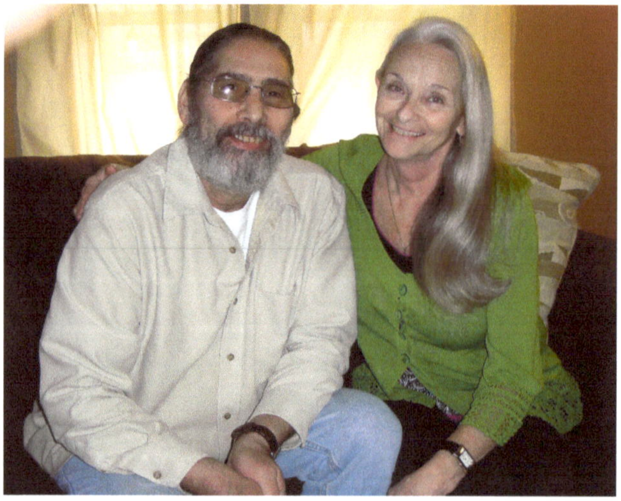

Acknowledgments

Words of wisdom and assistance from these family members made it possible to produce this book: my wife Colleen, daughters Colleen R. North, Debby North Kohn, Karen C. North, and son-in-law Richard Kohn.

"It takes a family to write a book."

Wm North

Bernadette Rashford, the "can-do" Director of Administration & Member Services of the Gulf Citrus Growers Association, Inc., for her invaluable help.

Jamie Kuser, *Happenings Magazine*, for his diligent photo research.

Megan Kissinger, whose creativity is evident on every page.

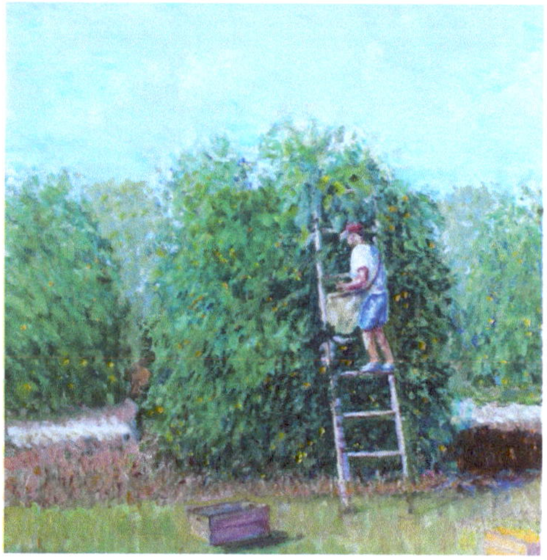

Dedication

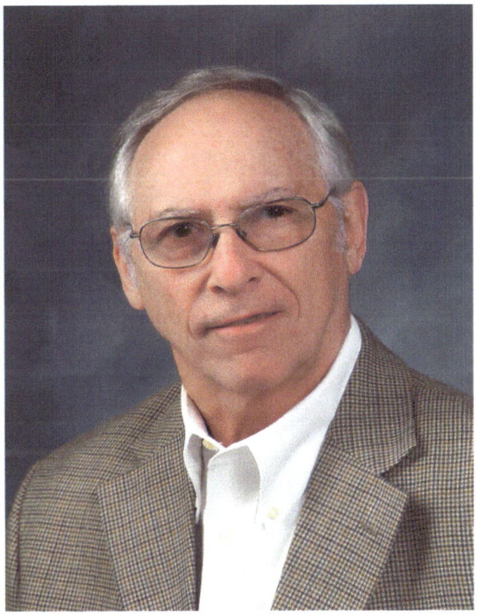

Ron Hamel

Executive Vice President and General Manager
Gulf Citrus Growers Association, Inc.
for his many years of encouragement and support.

and

*To all those who work in the citrus industry
to put juice in the glass and fruit on the table.*

Contents

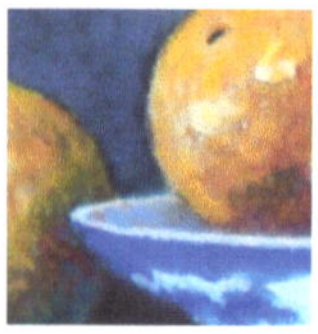

1 **Introduction**

Portrait of the Artist as a Father

1

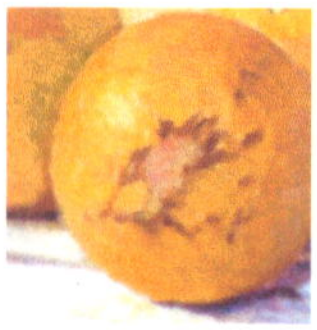

2 **Biography**

Highlights of the Artist's Life

5

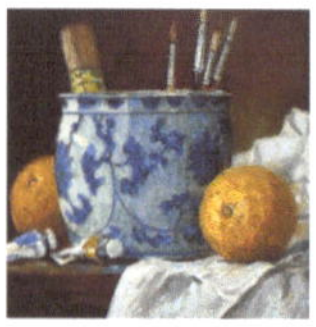

3 **One Artist's Vision of Florida Citrus**

The Story Behind the Paintings

9

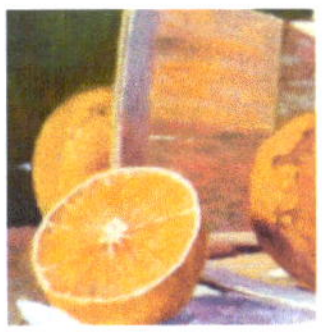

4 **Still Lifes**

Portraits of Florida Citrus

17

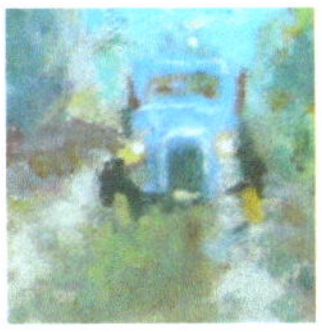

5 **Orange Groves**

Painting the Land of the Golden Harvest

71

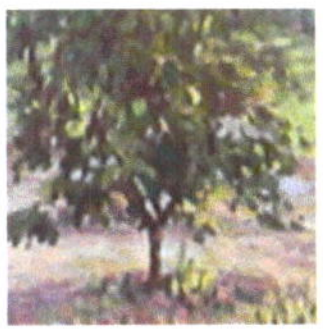

6 **Wm. North Gallery**

The Home of Florida Citrus Art

105

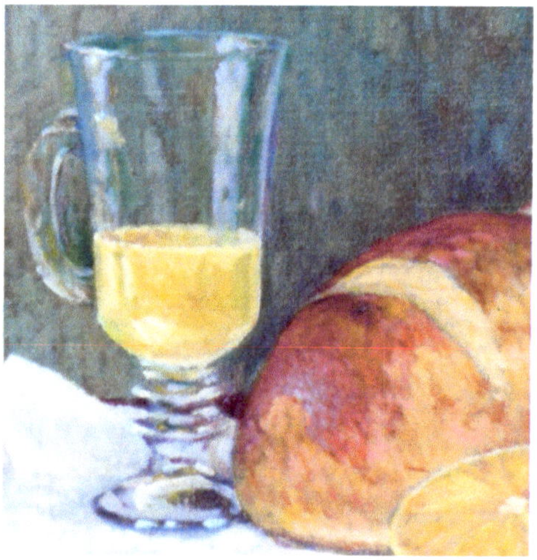

Introduction 1

Portrait of the Artist as a Father

Introduction

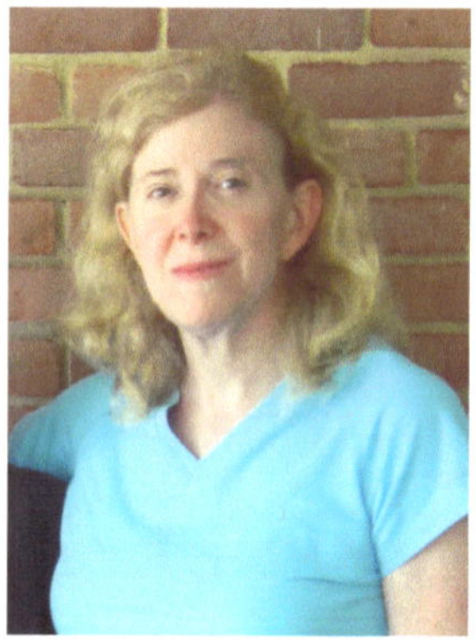

Portrait of the Artist as a Father

by Karen C. North

Art always has been part of my family. Not as an intellectual exercise, but as a backdrop that we were always conscious of. Everyone in the family was encouraged to do art of some kind. Wherever we traveled we always went to the art museums. We always stopped to smell the roses.

Family trips rarely went to mountains, deserts, or cities. We always seemed to end up near water and boats. We would go on fishing boats, ferries, historic ships, airboats, even ice boats. We walked and drove on beaches. Dad scuba dived and did endurance water skiing—once skiing in the ocean from Atlantic City, New Jersey, to Manhattan. Perhaps it was inevitable that he would end up in Florida surrounded by water.

My father has created many types of art over the years: oil paintings, of course, but also pastel drawings, pen and ink drawings, comics, photography, ships in bottles, models, a scratch-built runabout boat, and even a dinghy he named and painted to resemble a famous ship, *The Great Republic*. Even the yard needed to be mowed in a pattern to provide an artistic result! He kept busy, which is what it takes to turn talent into successful execution. Whether he was sitting on a dock painting fishing boats, or sitting on the beach painting sand, sea, and seagulls, or even in our backyard painting the "pond" left over from an abandoned gravel pit, it was all grist for the mill. During one trip to a lighthouse, he set up easels and palettes for us kids and had the whole family paint the lighthouse together. Even as an adult, I sometimes spent my vacations painting with him in Lee County, Florida. On two visits, I attended the annual Citrus EXPO with him and helped him to show his art.

One of his talents is to connect with the local themes. In New England, he found beauty in rocky coastlines and wooden piers. In New Jersey, he found beauty in fishing boats and sandy shores. In Florida, he found beauty in mangroves and citrus. There is nothing generic about his art. He paints impressionistically, but not abstractly. He paints scenes so that you can recognize their region. This helps him connect with local folk who seek art that shows off their locale and with tourists who want art that brings back fond memories of their travels.

His legacy will rest with his artwork and his family. His paintings speak for themselves. If Dad is disappointed that none of his children have become professional artists, he does not show it. He continues to imbue his children with an appreciation for art. Perhaps he waits patiently for them to have an epiphany and seize their opportunity, in turn, for a future in art.

My father is no ordinary man. He begins each day with energy, intensity, and creativity. This is how he approached his citrus art when, in a flash, he grasped its many artistic possibilities. The results can be seen on the following pages.

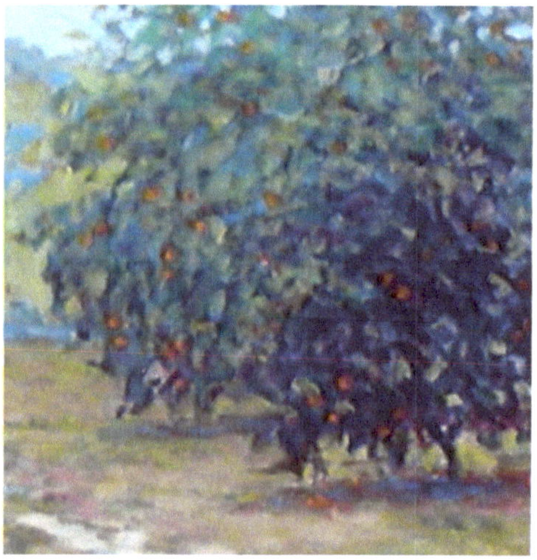

Biography 2

Highlights of the Artist's Life

Biography

Highlights of the Artist's Life

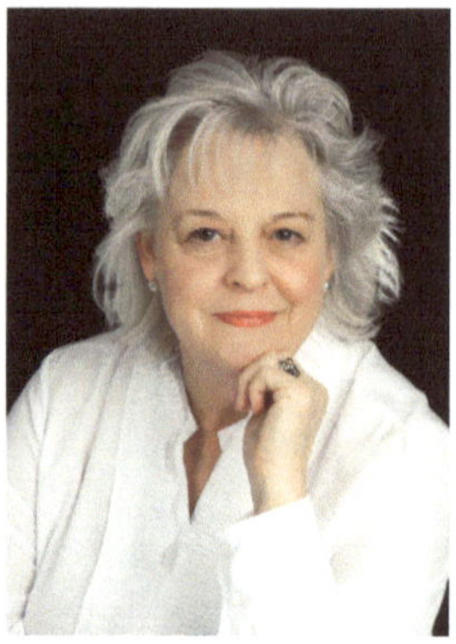

by Colleen R. North

William North is a well-known painter of the local scene, particularly noted for his portrayals of Florida oranges and orange groves.

An interest in art and oil painting, which began in his early teens and continued through high school, led to his assignment as an artist correspondent in the US Army for a GI newspaper in Germany (1945–47). There he honed his art and writing skills.

After leaving the service, he attended Pratt Institute in Brooklyn where he received his BFA in 1951. He continued to study at the Art Students League of New York and the Lyme Academy of Fine Arts in Connecticut. My father followed dual careers in art and business until 1990, when he left the business world and settled in Fort Myers to pursue his painting interests full time. He soon became established as an oil painter of local landscapes, completed outdoors in the impressionist tradition.

In 2004, he was the first Lee County artist to be invited to exhibit in the Governor's Gallery in the state Capitol

Complex in Tallahassee. Three years later, the Department of Environmental Protection and Delnor-Wiggins Pass State Park in Naples named him their first Resident Artist. That distinction came with appropriate honors, including a captain's chair labeled *Resident Artist* to use when painting in the park. My father has received numerous painting awards in his sixty-year career, twenty of which have been in Southwest Florida.

Living in the Sunshine State created an interest in citrus, which he translated into a large body of still lifes and grove scenes. Watching my father at his easel, I eventually absorbed enough of his methods and techniques to sit by his side and work on my own canvases. Spending studio time with my father, painting oranges in still life, produced paintings filled with cherished memories and treasured art to hang in my New Jersey home.

After experiencing problems with his vision, he was diagnosed in 2007 with macular degeneration, which along with glaucoma and cataracts, turned his world upside down. The progressive condition challenged him to find new ways to paint. He created several devices of his own and enrolled in the Veterans Affairs Low-Vision Program. The VA diagnosis of "catastrophically disabled" provided additional tools to aid him in his painting, including an enlarging Merlin Reader, which allows him to continue to read, write, and—most of all—paint.

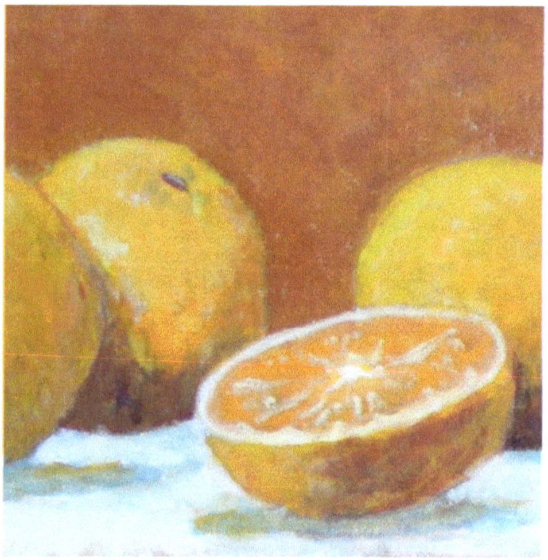

One Artist's Vision of Florida Citrus

3

The Story Behind the Paintings

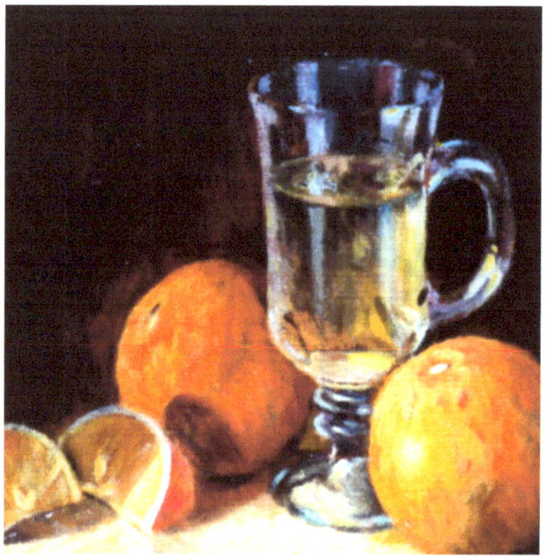

One Artist's Vision of Florida Citrus

The Story Behind the Paintings

1993 was an important year in my life...

It was the year I was introduced to the world of citrus, the year I did my first citrus painting, and the year my art took a new direction.

My wife Colleen and I moved to Florida in 1990, three years before my orange period. Immediately upon our arrival, I resumed my practice of painting out in the field to depict the local scenery. I was excited about the North-South differences: the tropic sky, the mangroves rooted in water, and the Sea Grape with its subtly colored leaves. The Palmetto and Cabbage Palms caught my eye because they were so unlike the popular image of a Coconut Palm. The southern landscape interested me, and I wanted to capture all of these

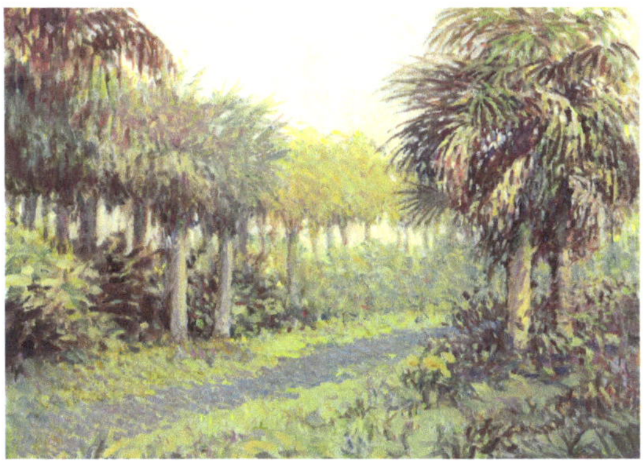

tropical differences and the feeling of Florida sunshine on my canvas. In three years, I produced a large body of work, most of which was sold in local galleries.

I painted in the cool of the morning and, after four or five days, took the nearly completed canvas to my studio where I added the final touches and signed my name. This process left many empty hours so I decided to explore the possibilities of doing still lifes in the afternoons.

In search of subject matter, I went to a nearby produce stand, the Cabbage Patch, on US Route 41 in Fort Myers to see what I could find. The moment I stepped inside the door I was struck by the sight of a mound of oranges. I took a closer look and saw, for the first time, real Florida oranges. They were unlike any I had seen before. Not "pretty as a picture" like those in a supermarket or as portrayed on juice cartons. Their skin was covered with blemishes. Many had scratches or spots, and some had large shiny areas that looked like islands or continents. They were multicolored, with touches of brown, red, yellow, and green mixed with the orange color.

I spoke to the proprietor, Jon Speigel, who explained that the large blemishes were wind scars caused by a twig or leaf brushing against the fruit as it matured. He said, "No matter what the outside looks like, Florida juice oranges are the sweetest in the world." Though the

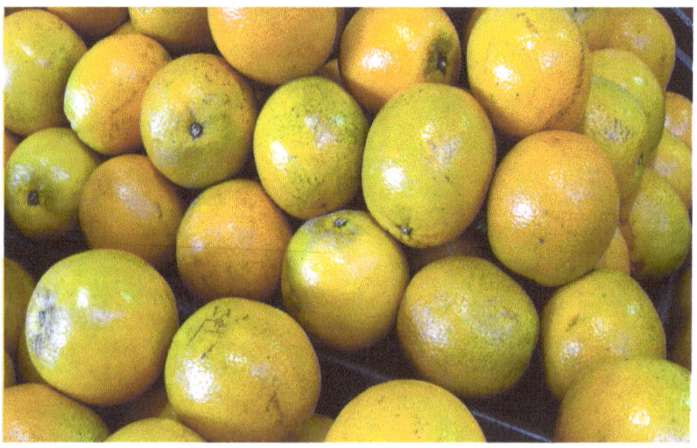

appearance of the oranges may have disappointed the growers, to the artist's eye they were pure gold. I had found my still life subject; I would paint the oranges as they really are. With a bag of Valencias purchased at the Cabbage Patch, I returned to my studio and arranged them in a traditional still life setting. I began my *Real Florida Oranges* series, which continues to this day.

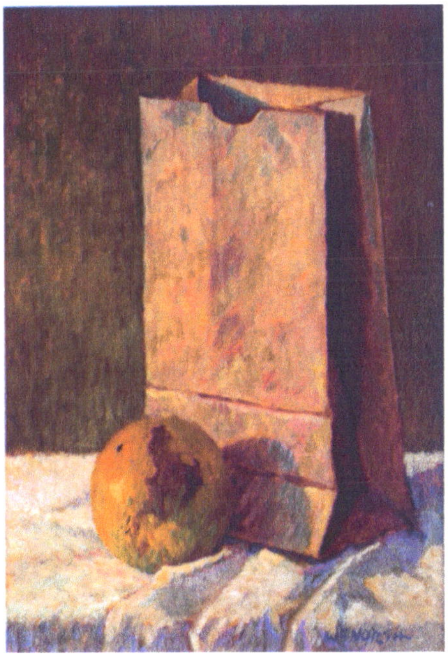

For my first effort, I placed the fruit in a clear glass bowl on a draped, cream-colored cloth. The oranges above the rim of the bowl were painted realistically—as they appeared. Below the rim, the glass distorted the images for an interesting, artistic effect. I then moved on to fluted glass, brass containers, oriental pots, and a great variety of other accessories. Each orange was a portrait of a singular juice orange, just as if it were a portrait of an individual person.

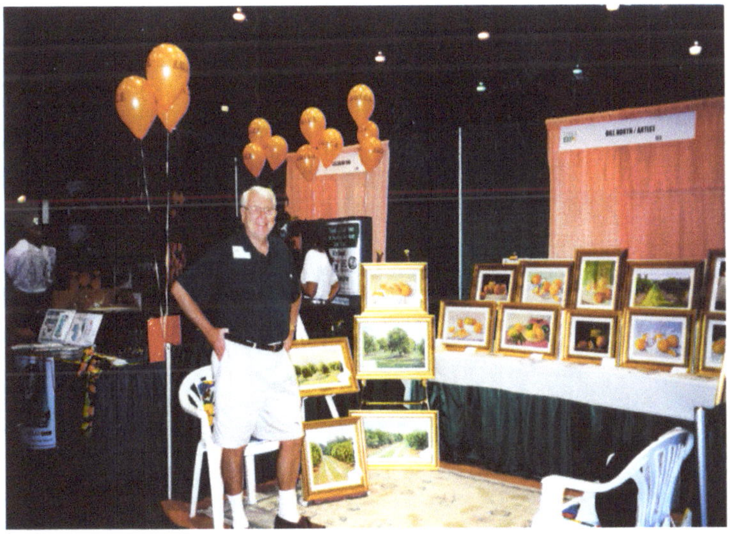

Twelve years ago, because of my interest in citrus, I became aware of the Citrus EXPO trade show held in the Lee County Civic Center. I took a booth at the EXPO to display and sell my art. Here my paintings were shown among the John Deere and Ford tractors, mechanical harvesters, Tropicana, Florida's Natural and Southern Gardens processors, and many other providers of goods and services to the industry. The result of my participation was a gratifying success. In addition to making art sales, the compliments flew. Many growers approached me to say that they had never before seen Florida oranges pictured as they really are.

My orange still lifes led to a logical interest in seeing and painting citrus groves. The next time I was at the Cabbage Patch, I mentioned that I would like to paint them to the owner Jon Spiegel. He perked right up and said, "Let's go!" We hopped into his pickup and drove about a mile away to his fifty-acre grove just a stone's throw from busy Route 41. I did my first grove painting there. It was a great site; a large oak tree provided shade, and there were a variety of different views to paint in every direction.

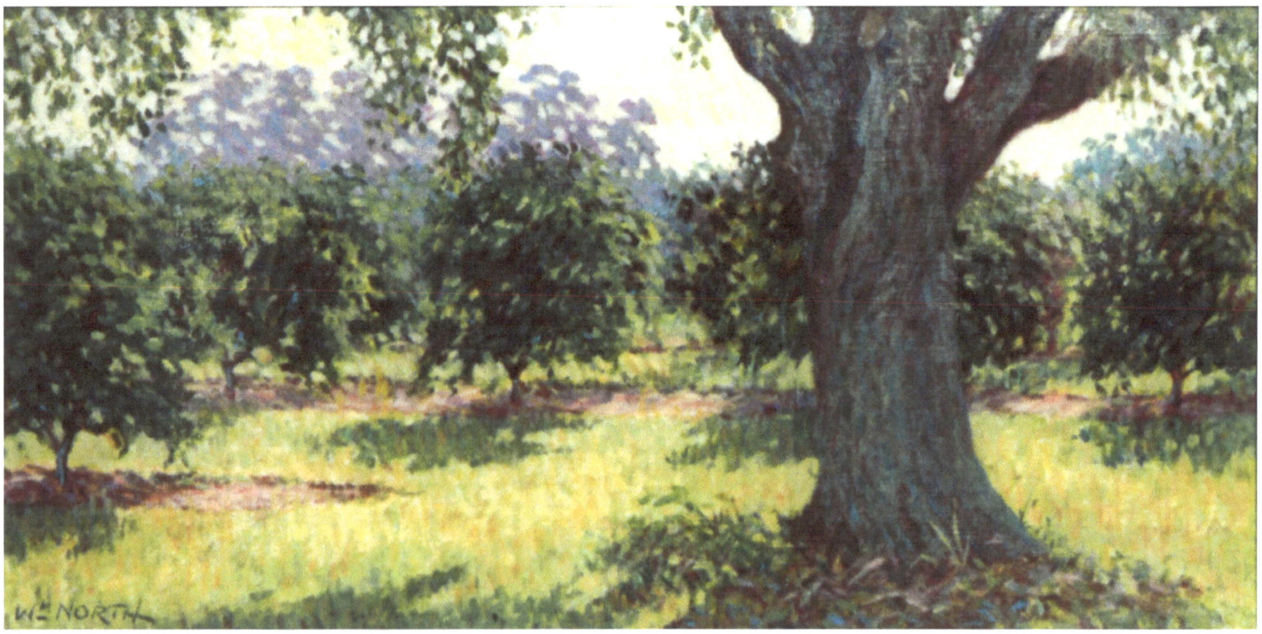

The Shade Tree

Painting in this location for several years, I had a constant companion. He was a neighbor's large, black dog. After a brief get-acquainted period, he attached himself to me, and every morning he would wait for my arrival, greet me, and stand by as I unloaded the car and set up my easel. He then would lie down in a shallow depression that he dug in the sandy soil and nap the morning away. When I moved to a slightly different spot for a new painting, he would scratch out a new depression under my easel and pick up his nap from the day before. Our friendship ended when I didn't show up one day, having moved to a new location at a George Austin grove in Alva.

In the following years, other generous owners allowed me to work on their properties so I could get a fresh perspective. As time went on, I added new elements to the subject such as orange crates, vehicles, birds, and people. Wooden crates are no longer used, but I sometimes include them to give the paintings a nostalgic flavor.

As of 2011, I've completed eighteen years of citrus paintings, and I'm still at it. Hundreds of my paintings have been purchased and are in the homes and offices of collectors wherever citrus is grown. Some of the paintings and the stories behind them are on the following pages.

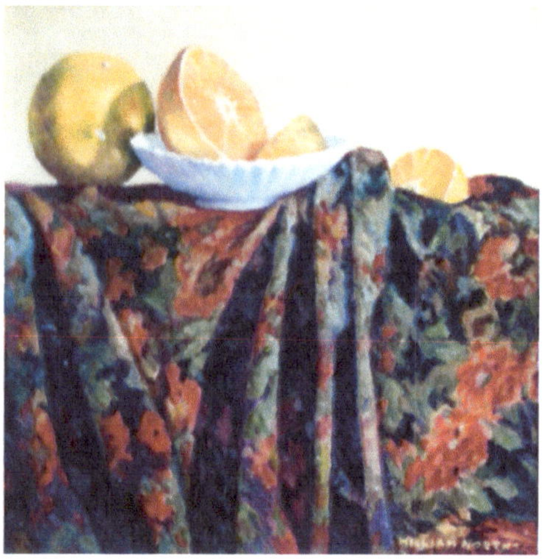

Still Lifes 4

Portraits of Florida Citrus

The "Command and Control" Center

Painting a still life in my studio's "Command and Control" center.

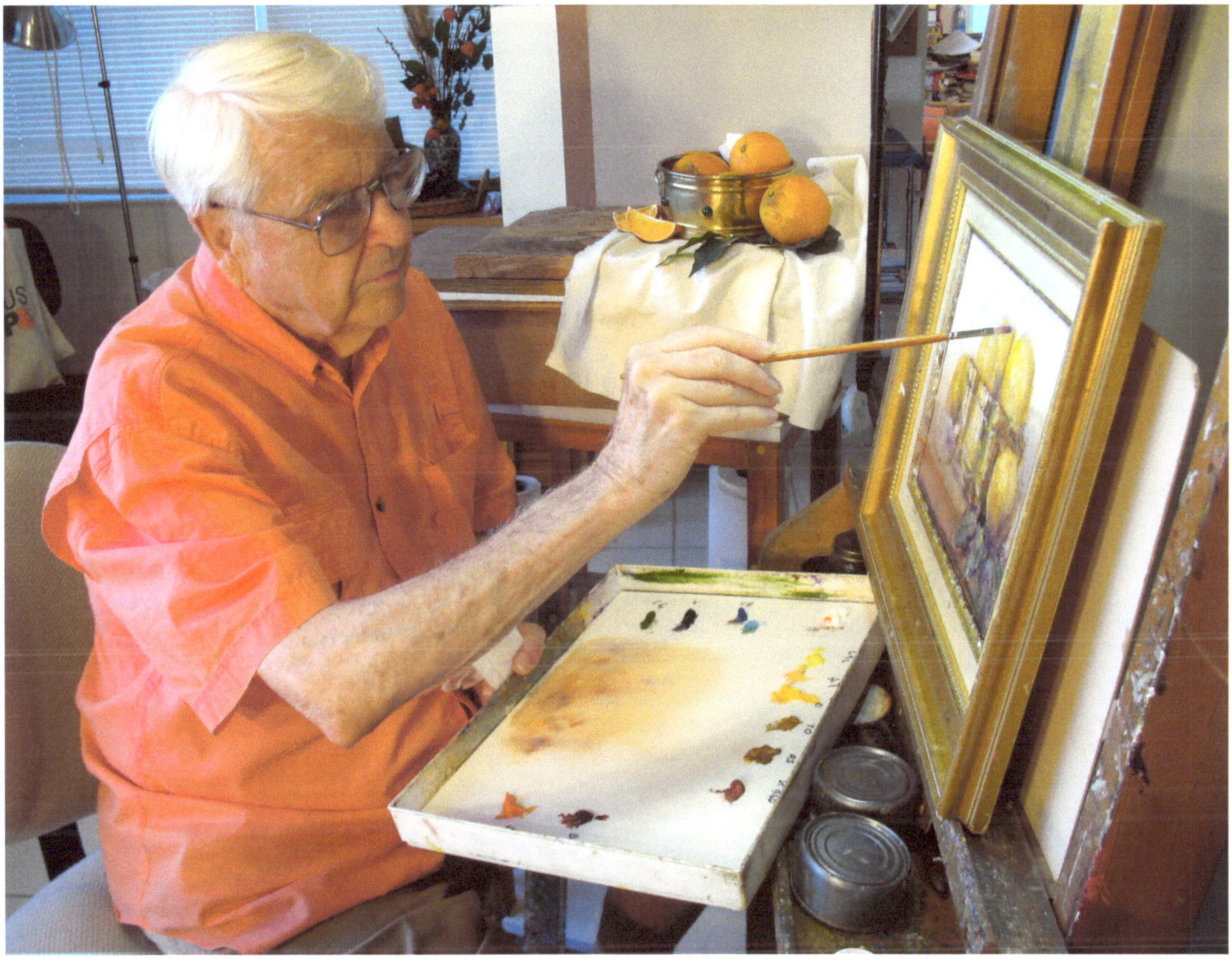

Valencias in Clear Glass Bowl

Simple and direct, the classic simplicity of the clear glass bowl allows the brilliance of the oranges to show through. Eventually, I'd use a variety of other glass containers in my paintings, like fluted and cut glass, but I think this reflects the "pure gold" I first saw at the produce stand.

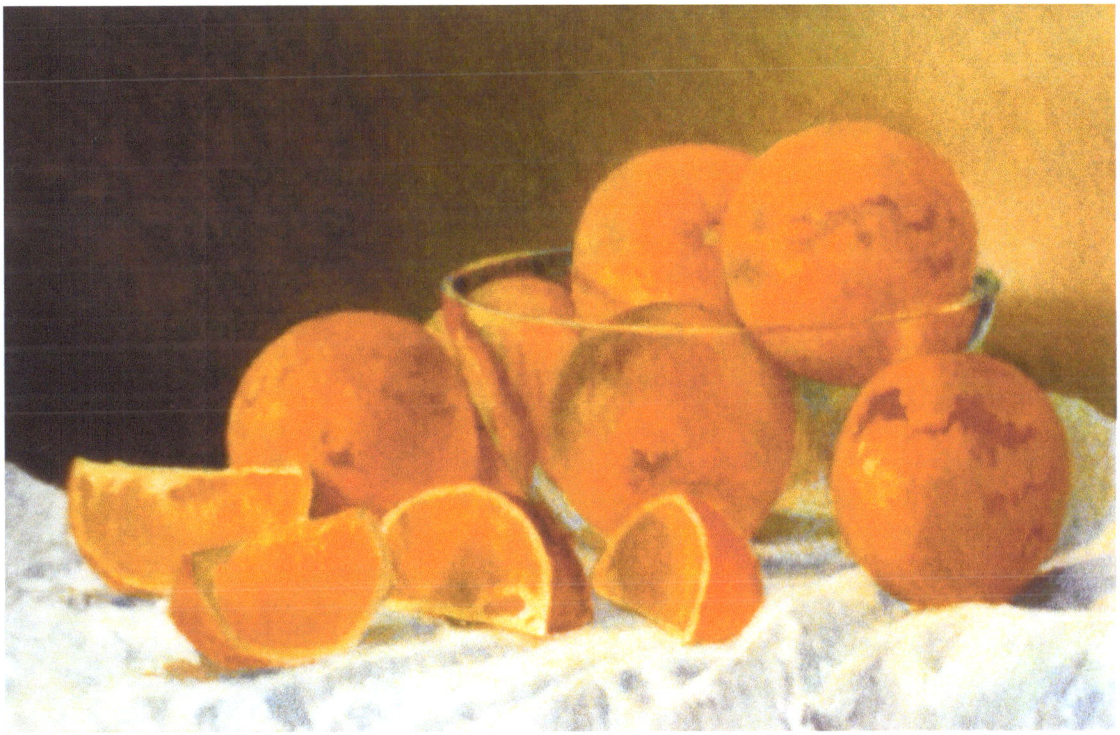

VALENCIAS IN CLEAR GLASS BOWL

Fluted Glass Bowl

These real Florida oranges command our attention, showing off their abstract reflections from inside the fluted glass bowl. The simple folds in the tablecloth and sedate background help to focus our eyes on the color and vibrancy of the fruit. My artist friends complain about the difficulty of painting glass, and my answer is, "It's really simple; you don't paint the glass, just paint what you see through it."

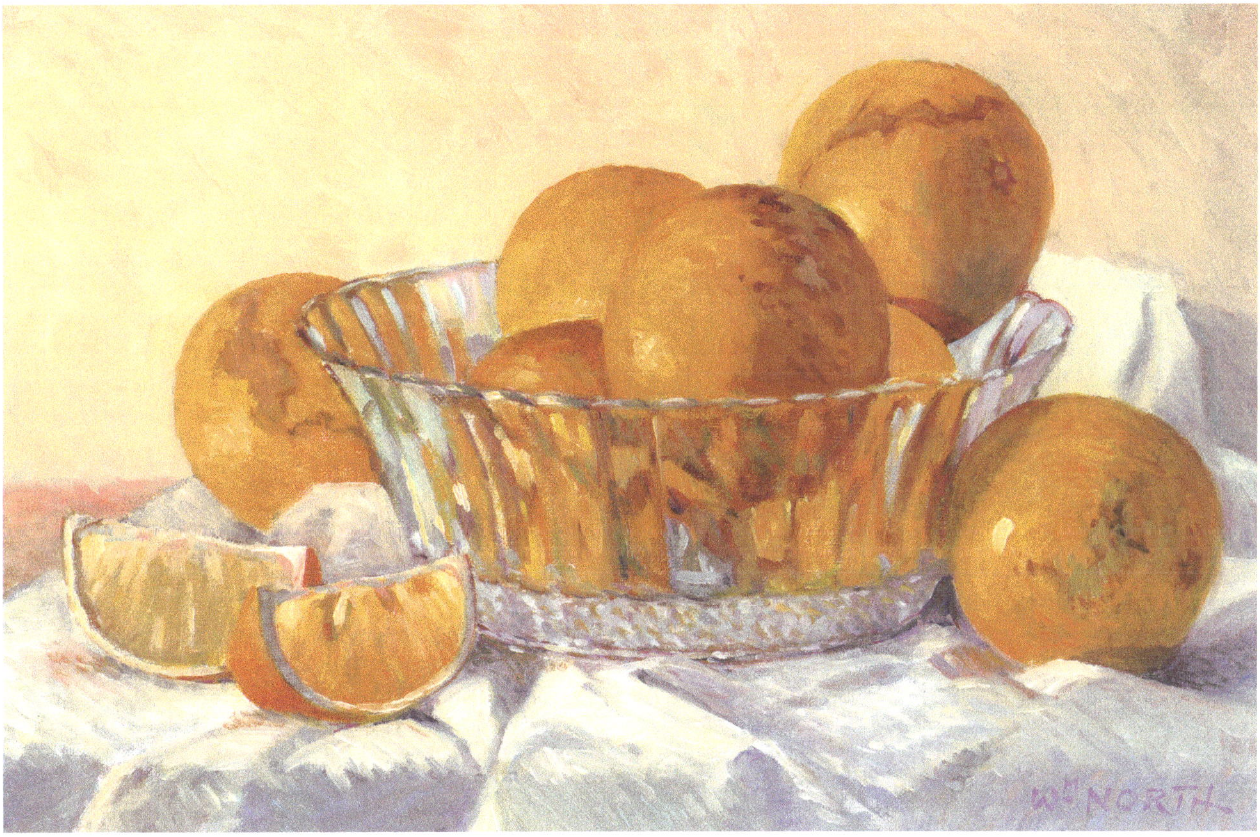

FLUTED GLASS BOWL

Red Valencias in Brass Bowl

Ready to eat, or just pretty to look at, the golden glow of sunshine permeates this painting. I picked up the container for $3.99 and knew that real Florida oranges would shine when placed in it. The bowl was made in India, probably from World War II shell casings.

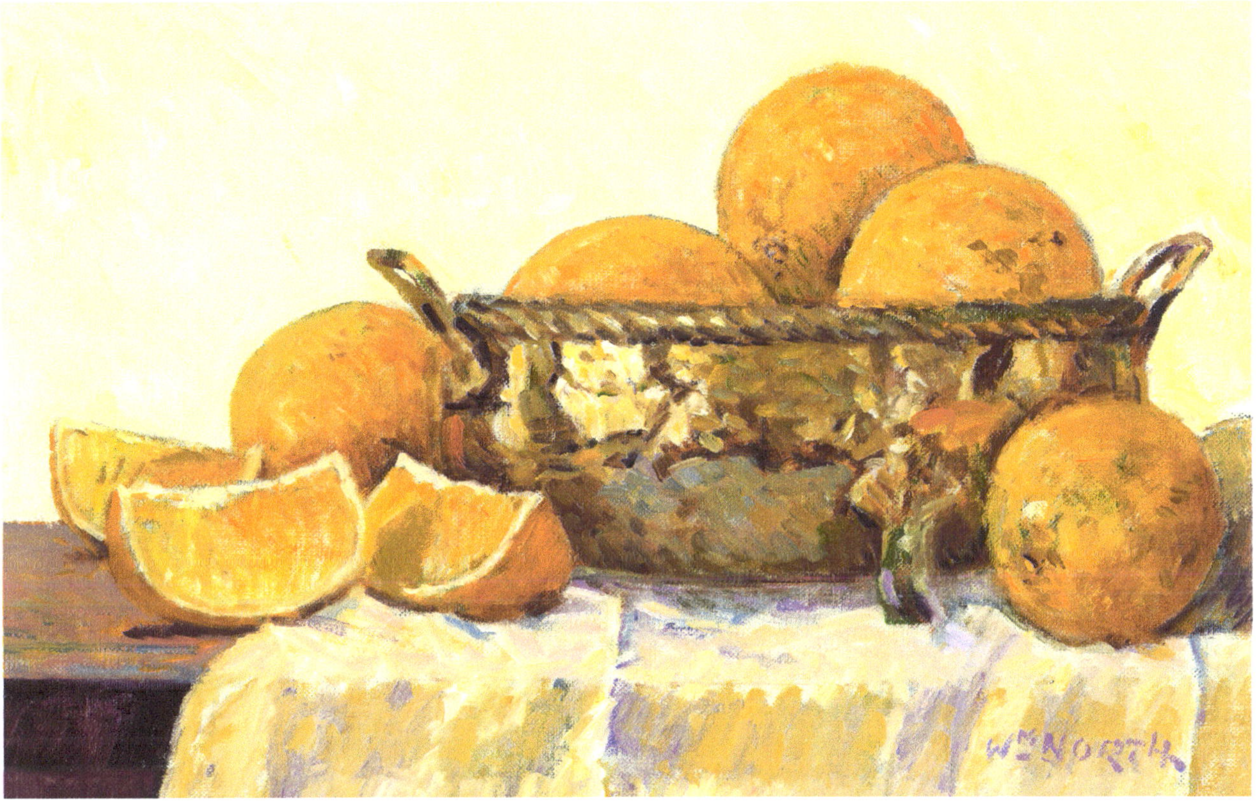

RED VALENCIAS IN BRASS BOWL

Artistic License

Green leaves play an important role in this painting. They're not real citrus leaves, but artificial ones I've used as props in other paintings. They break up the orange color by their size and dominance.

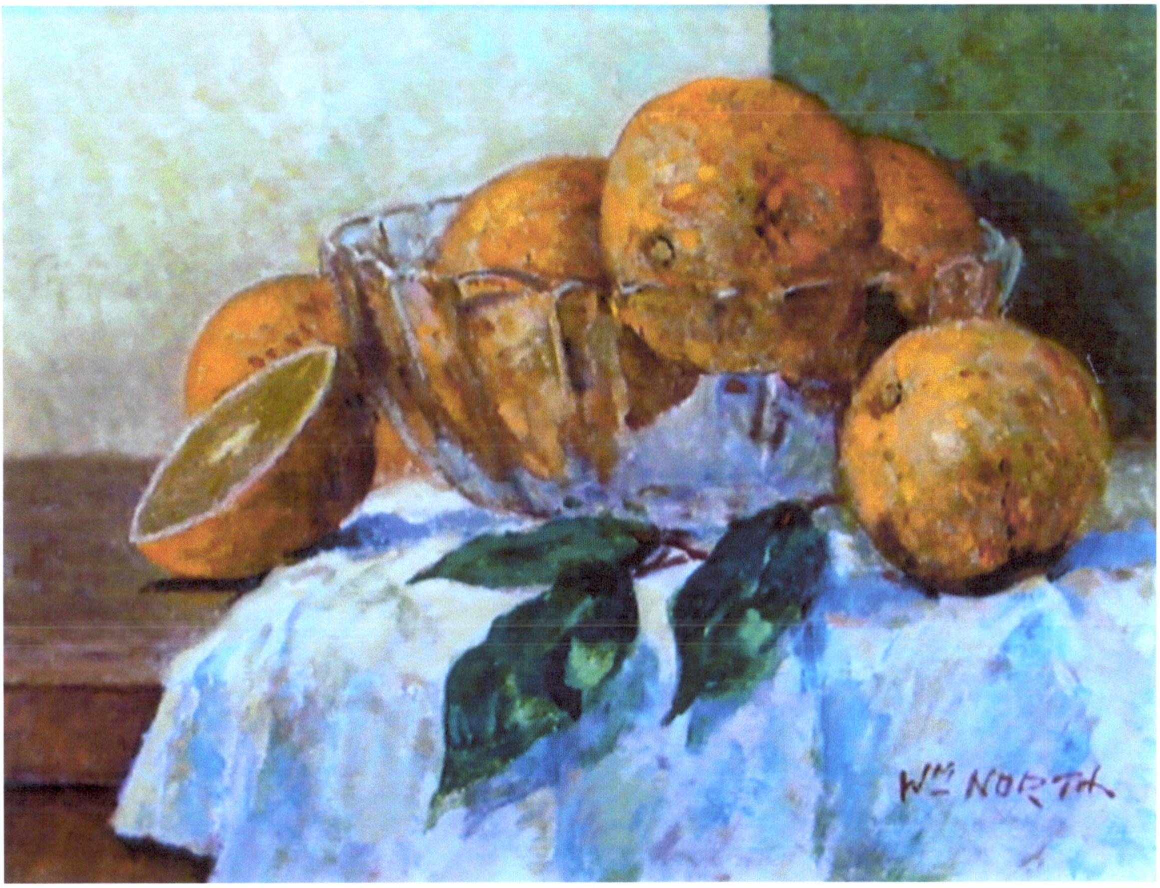

ARTISTIC LICENSE

Florida Oranges, Chicago Rolls

The title came to me while thinking about how many visitors to my gallery come from Chicago. When Chicago tourists visit, I always show them a print of this painting. Though Publix Supermarket sells oranges and these rolls, many real Chicagoans tell me they've never heard of Chicago rolls.

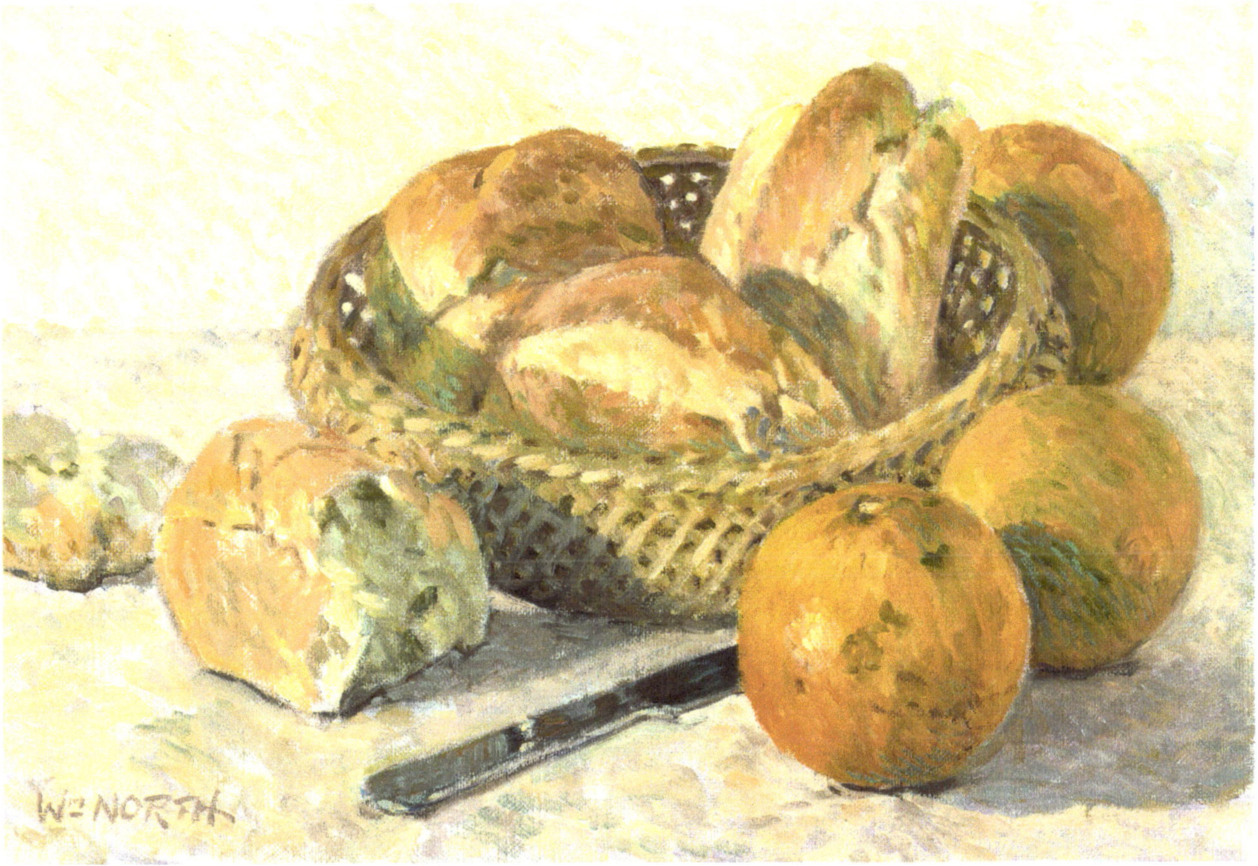

FLORIDA ORANGES, CHICAGO ROLLS

Florida Oranges in the Artist's Studio

An oriental blue vase holding my brushes, well-used tubes of oil paint resting on the table, and oranges tie the painting together. To me, this composition represents my most significant blending of citrus and art. Historical footnote: The fruit is believed to have originated in China, and the use of the oriental bowl pays homage to this. Curiously, the first known written word about the orange is in a text edited by Confucius.

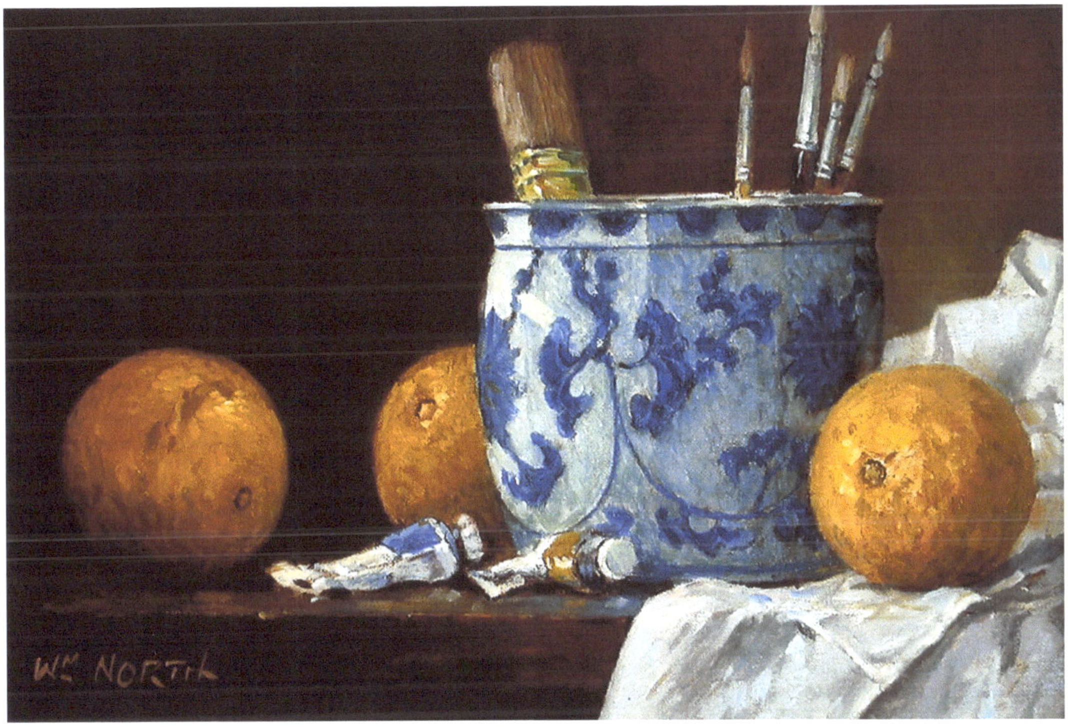

FLORIDA ORANGES IN THE ARTIST'S STUDIO

Valencias with Wind Scars

My eyes popped when I saw these oranges on the produce rack. They had everything I wanted—lots of character and potential for a painting. When I put them in my still life shadow box and added a small sprig of leaves, I knew they could stand alone. No bowl or props were needed; the subject was wind scars.

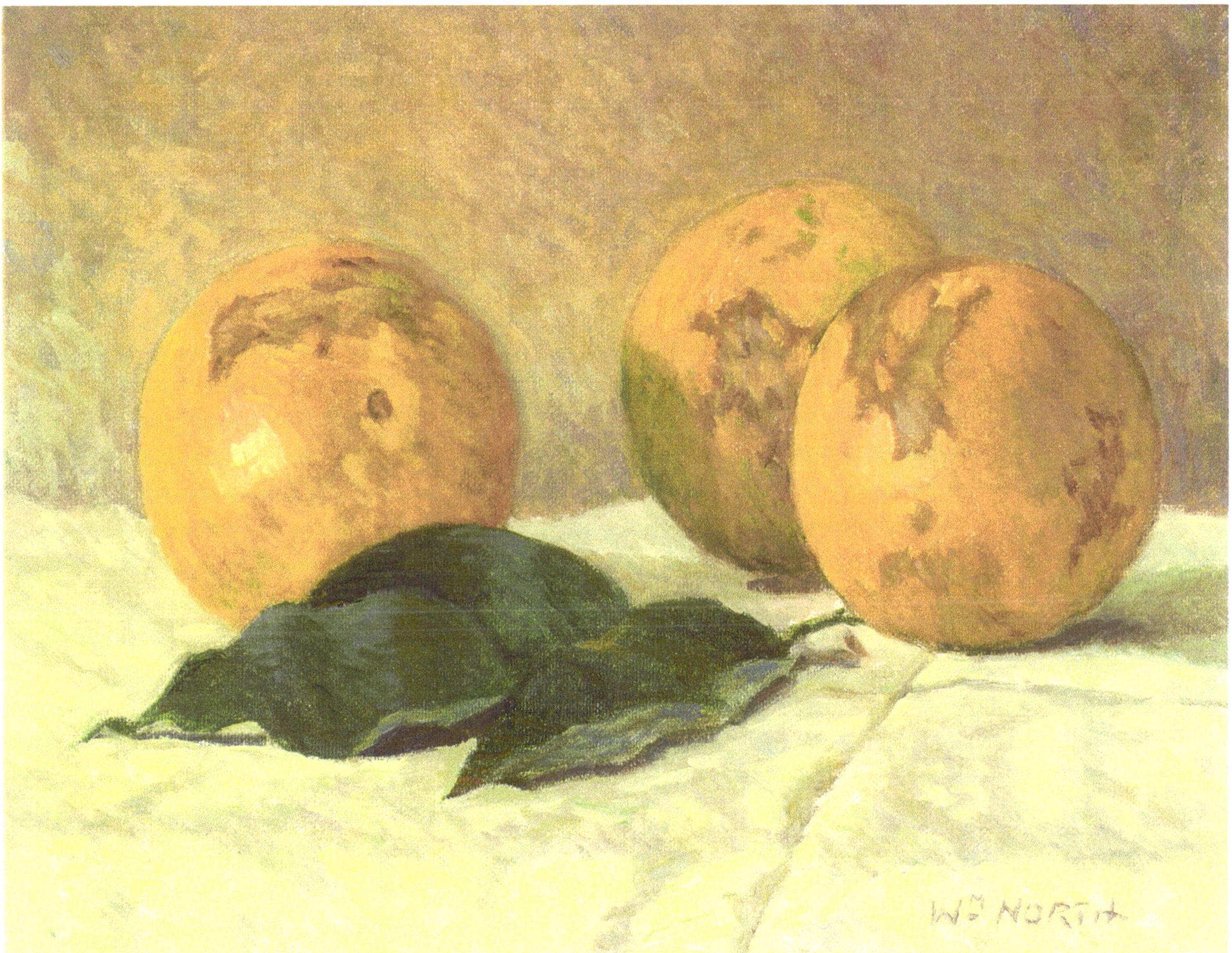

VALENCIAS WITH WIND SCARS

Valencia Reflections

Dark paintings like this one are a departure from my norm. Usually I am trying to convey the feeling of Florida sunshine in my work, but the opportunity to render the reflections in the brass cup was intriguing. A surprising number of my patrons prefer a dark picture; I think they like the classical feel. Friends refer to this as my Rembrandt period. The original was sold long ago.

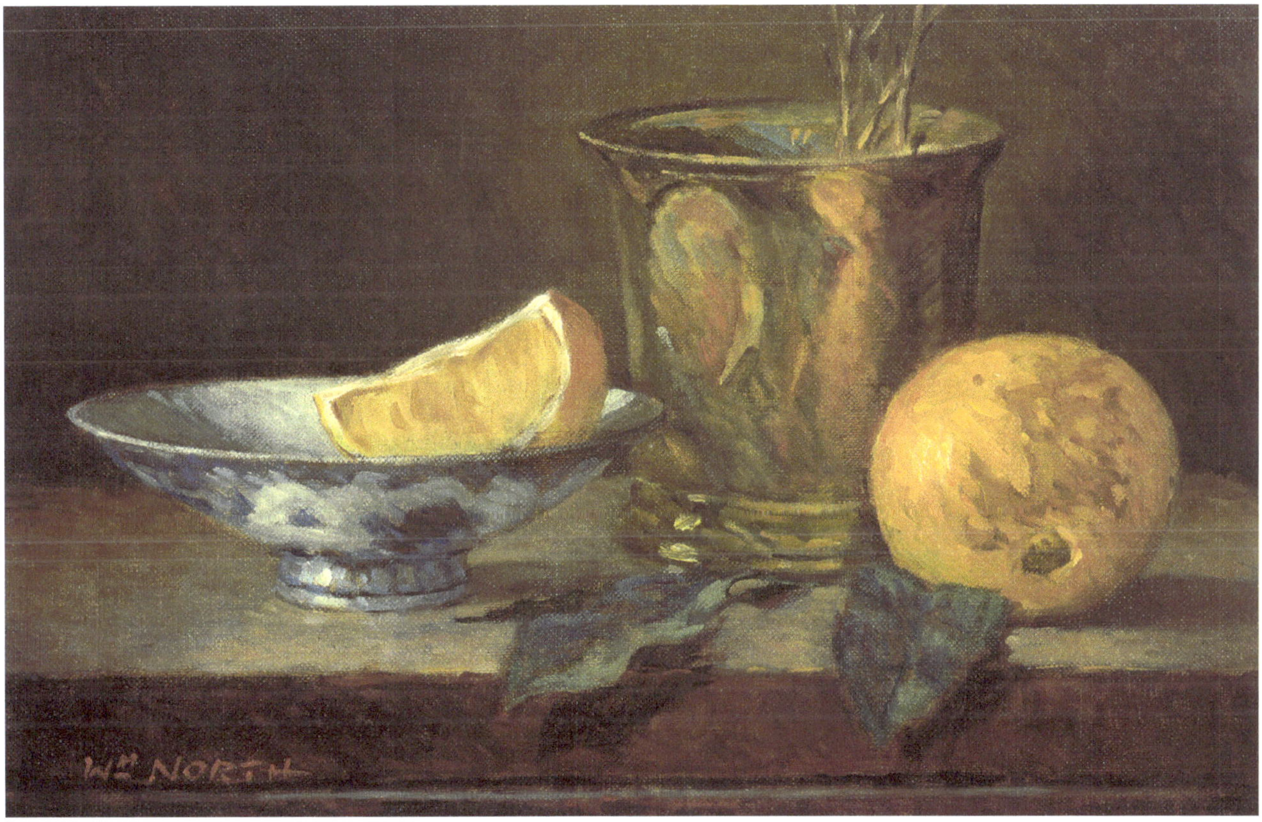

VALENCIA REFLECTIONS

Five Florida Valencias

This painting was used as the title illustration by a Sanibel Island magazine for a story about the history of the orange industry in Florida.

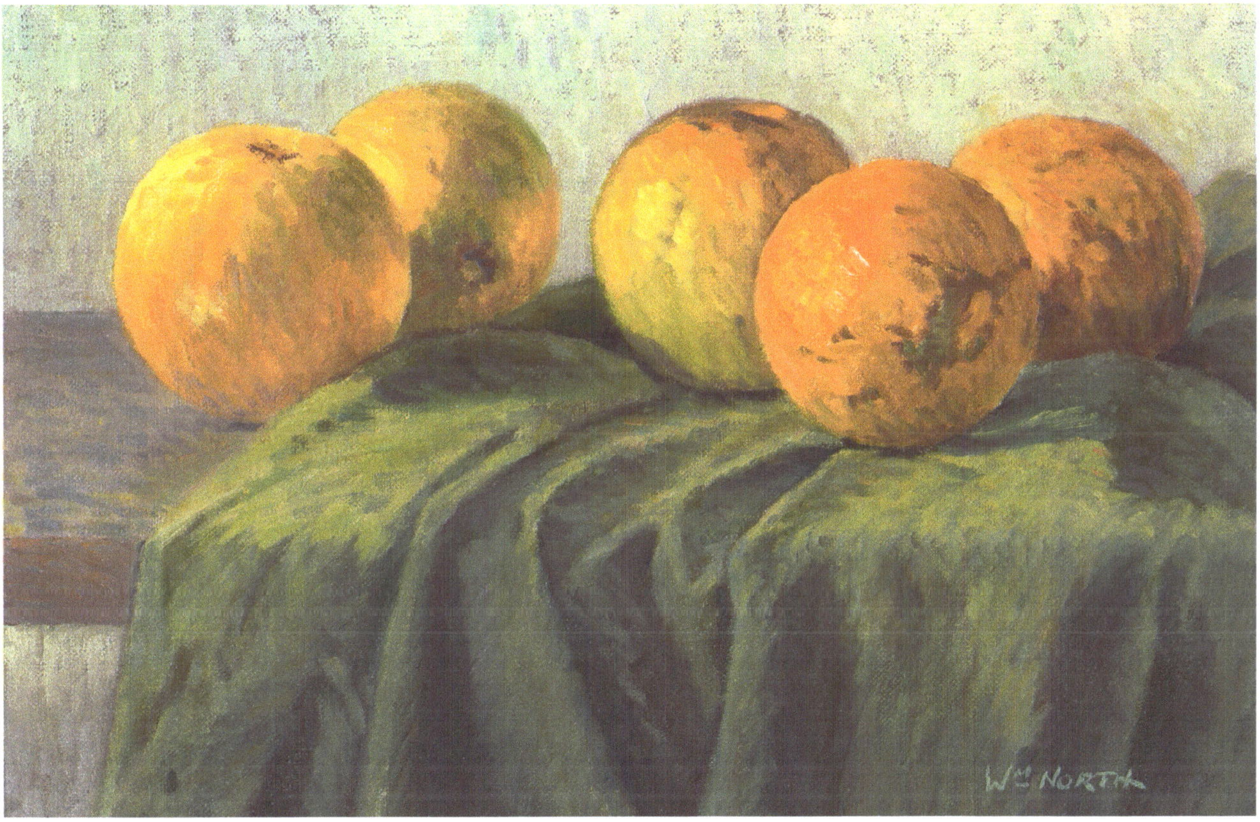

FIVE FLORIDA VALENCIAS

Now Showing

With its simple scalloped rim and slightly fluted sides, this glass bowl was an ideal container that allowed the oranges to perform at their best. It is the oranges, of course, that are the stars in this show.

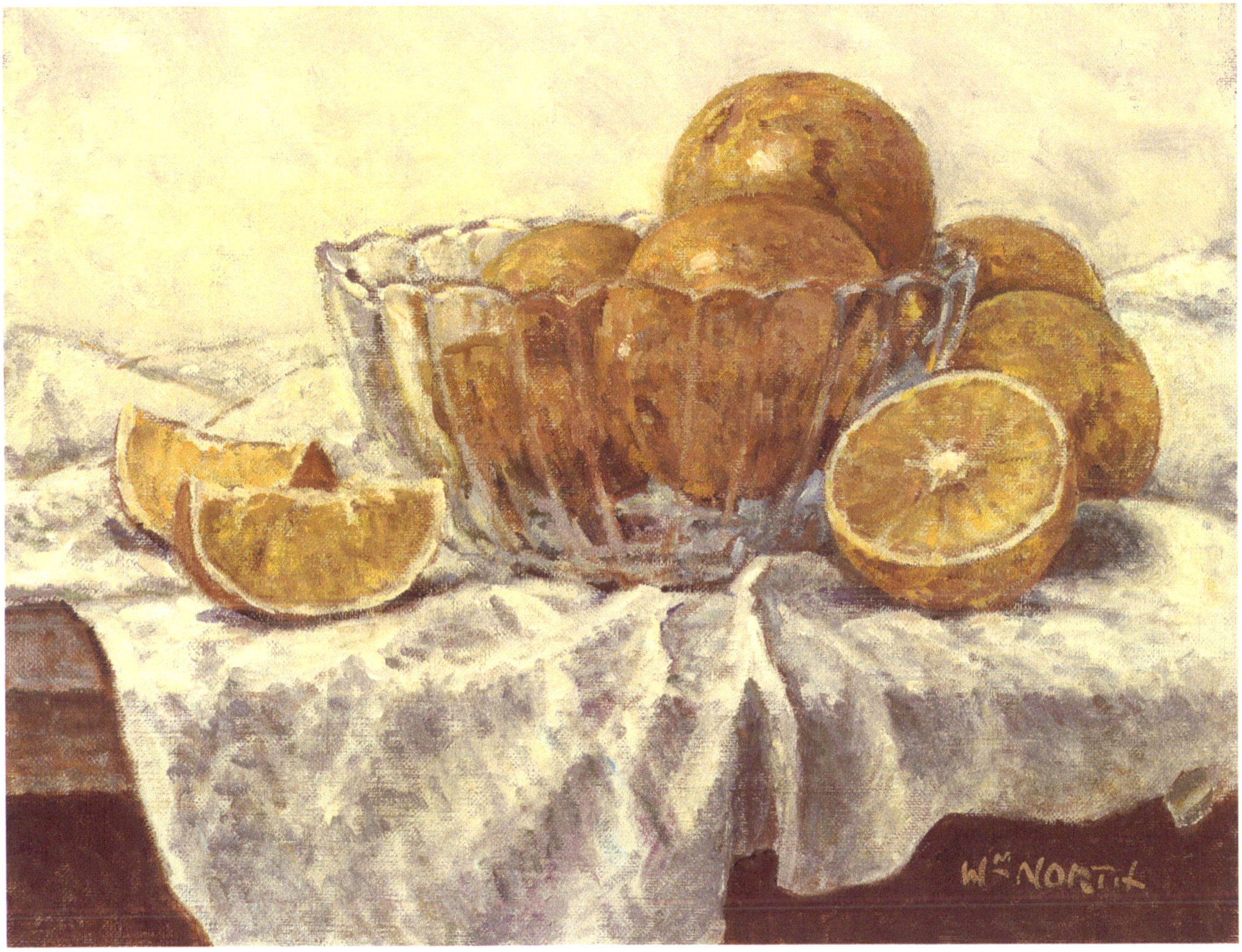

NOW SHOWING

My Gallery Made Me Do It

If I had my way, my paintings would have no titles. However, throughout years of exhibiting in commercial galleries, their proprietors have insisted that I title my work. So I called this piece "Produce Box With Oranges," meanwhile believing the viewer can clearly see that this is a produce container with oranges.

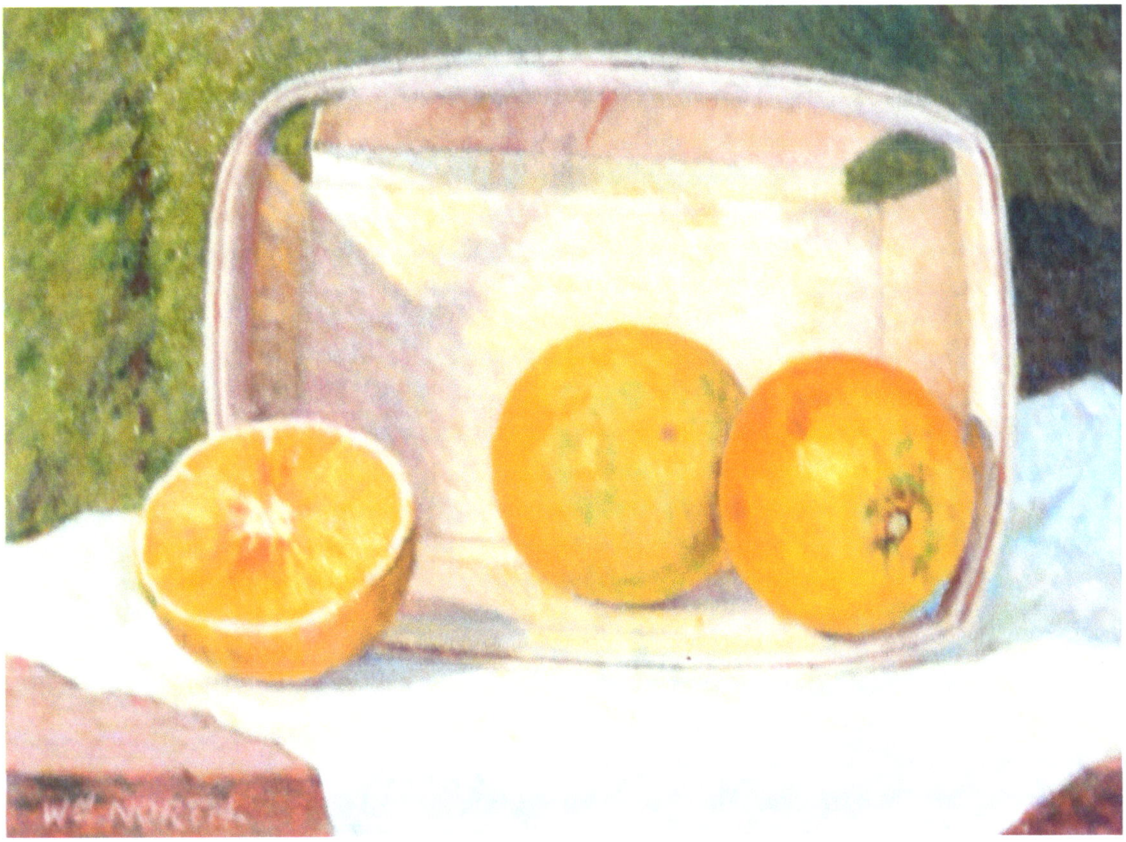

MY GALLERY MADE ME DO IT

I Did It Because My Gallery Said So

Ditto!

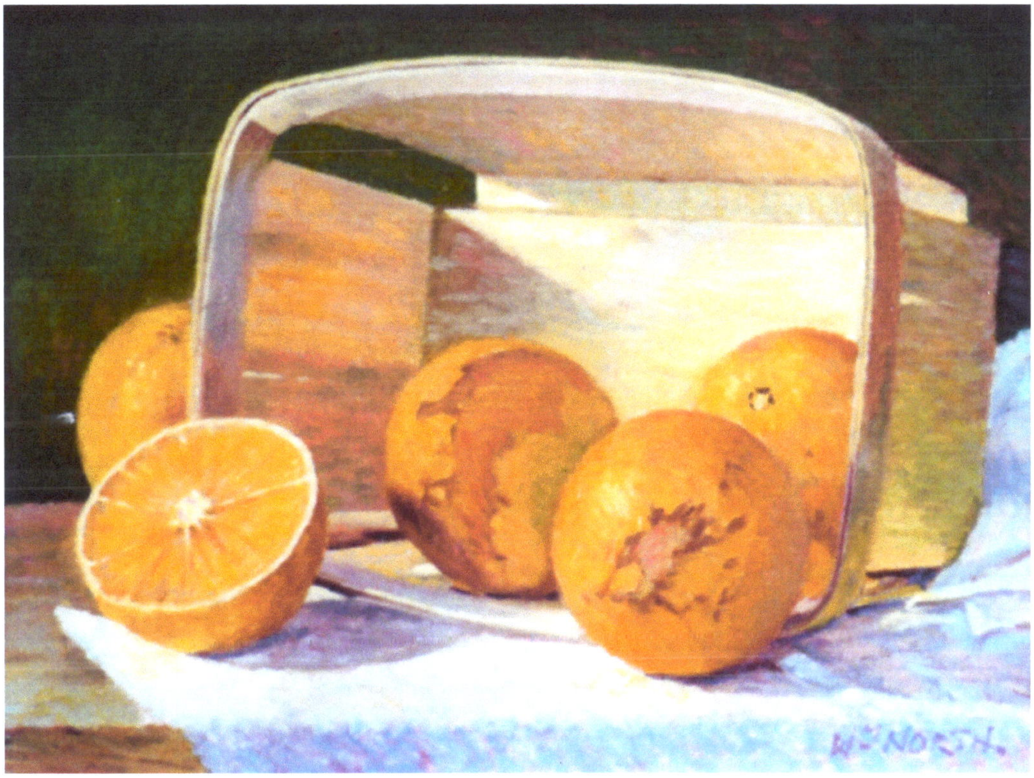

I DID IT BECAUSE MY GALLERY SAID SO

Marmalade

The round marmalade jar peeking from behind the bread can be found in tourist stores throughout Florida. After completing this painting, I sent a photo and note to the Davidson of Dundee company in Dundee, Florida, and offered it for sale. I didn't get a direct response but, instead, received a box of jellied citrus fruit candy. I think it was their way of sweetening the rejection.

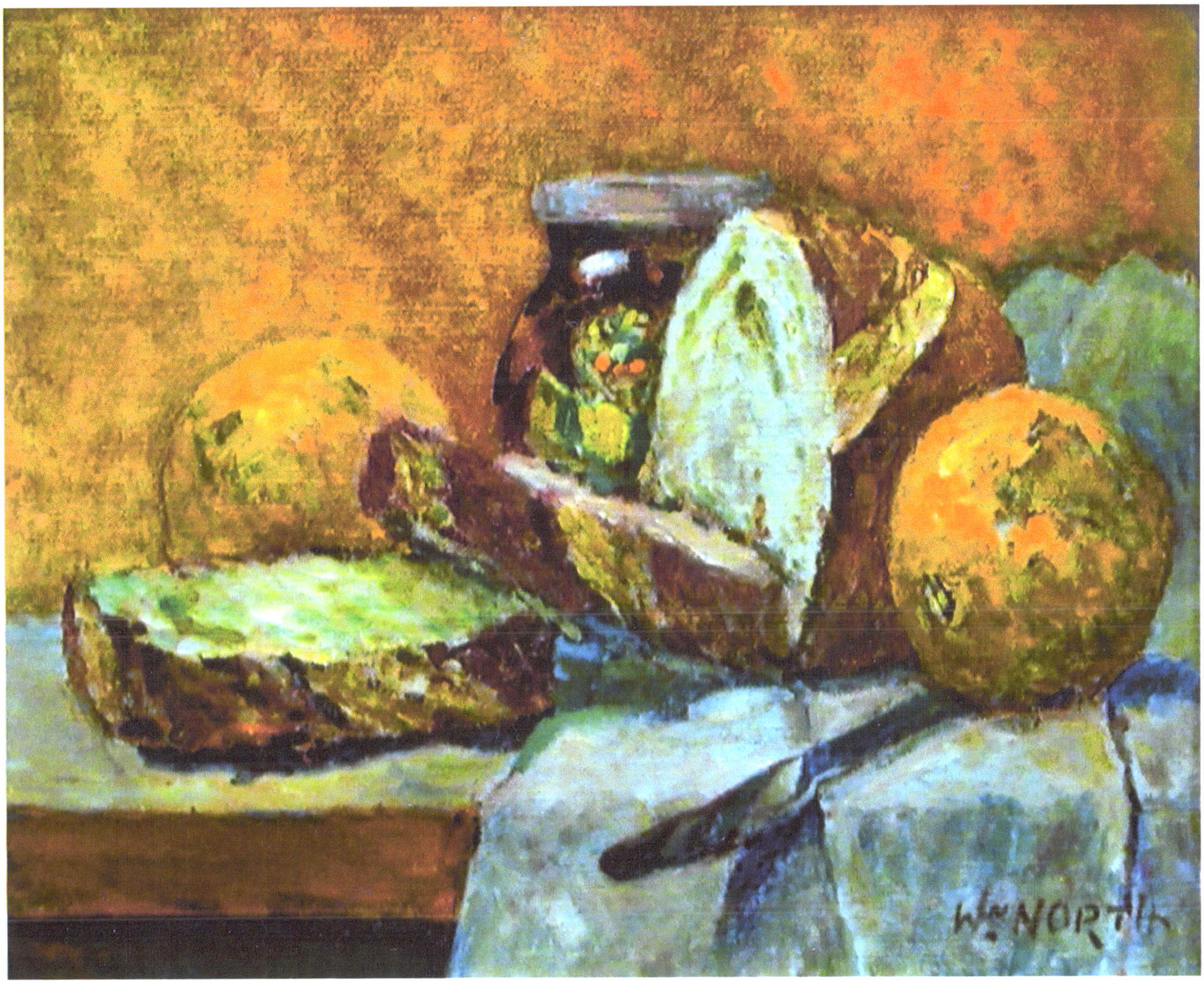

MARMALADE

Oriental Soup Bowl

Oranges from the produce stand were arranged around and in this oriental soup bowl to take advantage of the near complimentary color enhancement.

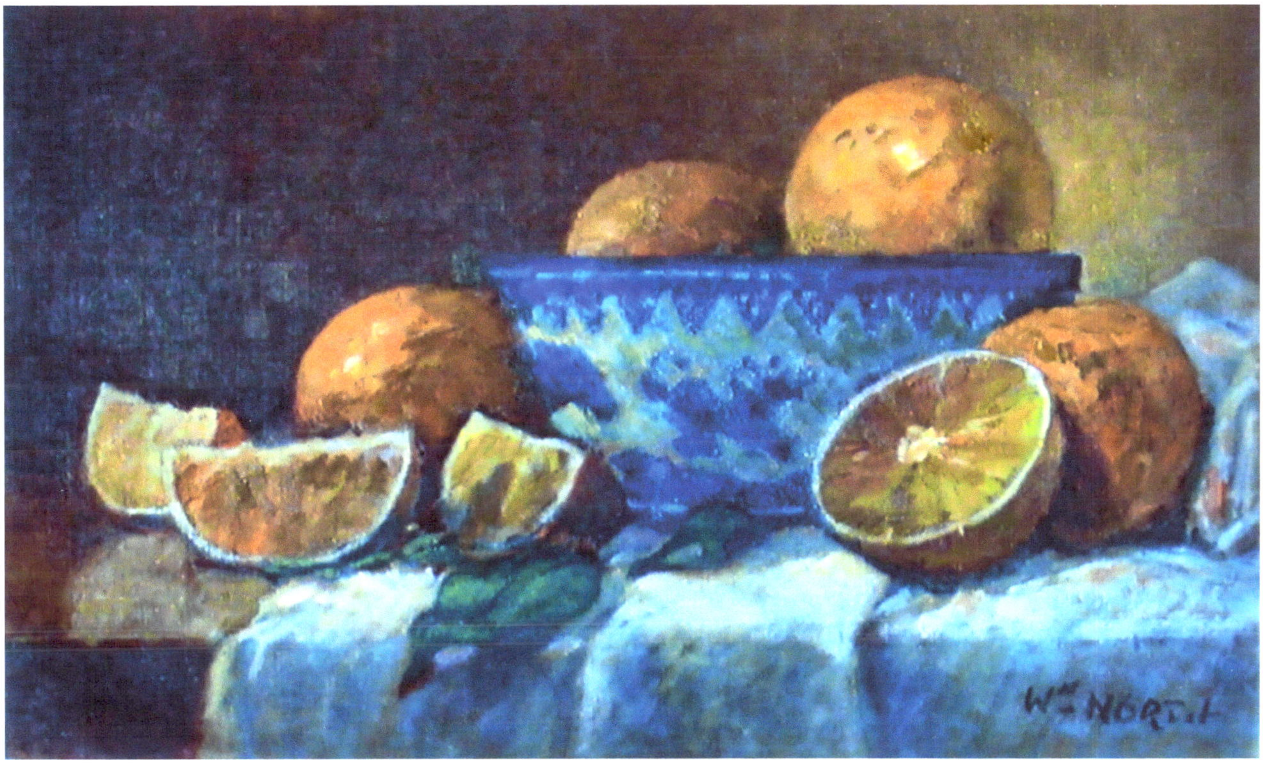

ORIENTAL SOUP BOWL

Oranges on the Table

Historians tell us there were no oranges in the Holy Land during the time of Christ. When Leonardo DaVinci painted oranges on the "Last Supper" table, they must have been Renaissance era oranges.

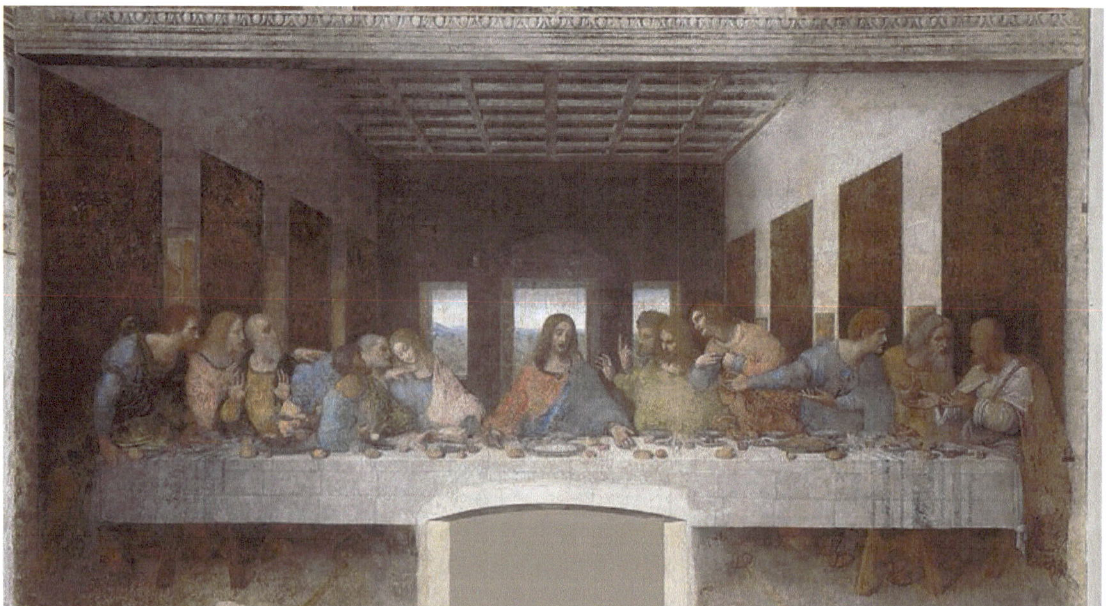

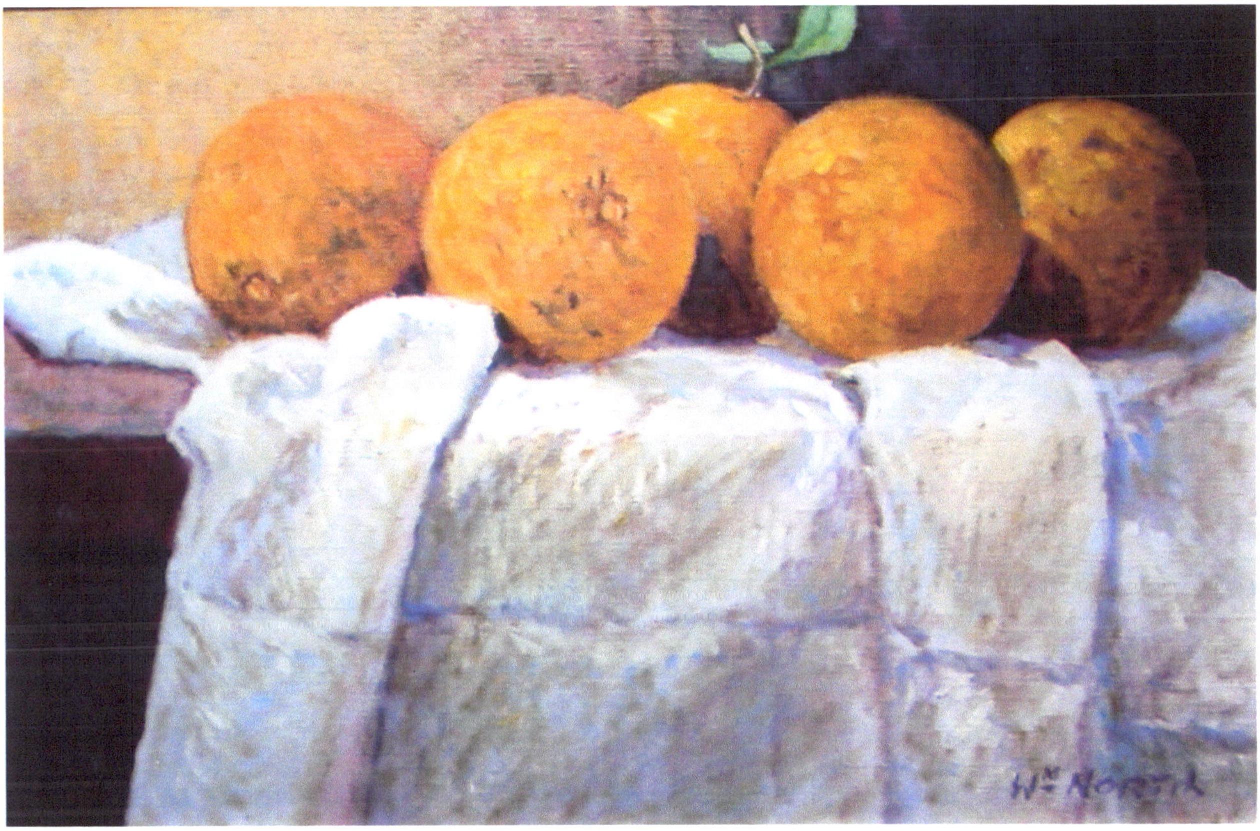

ORANGES ON THE TABLE

Bold and Fresh

Loose brushwork and bold colors are the outstanding characteristics of this work. As loose as it is, it still maintains all its form in a precise manner. The container, in particular, has a wealth of fresh color in its reflections.

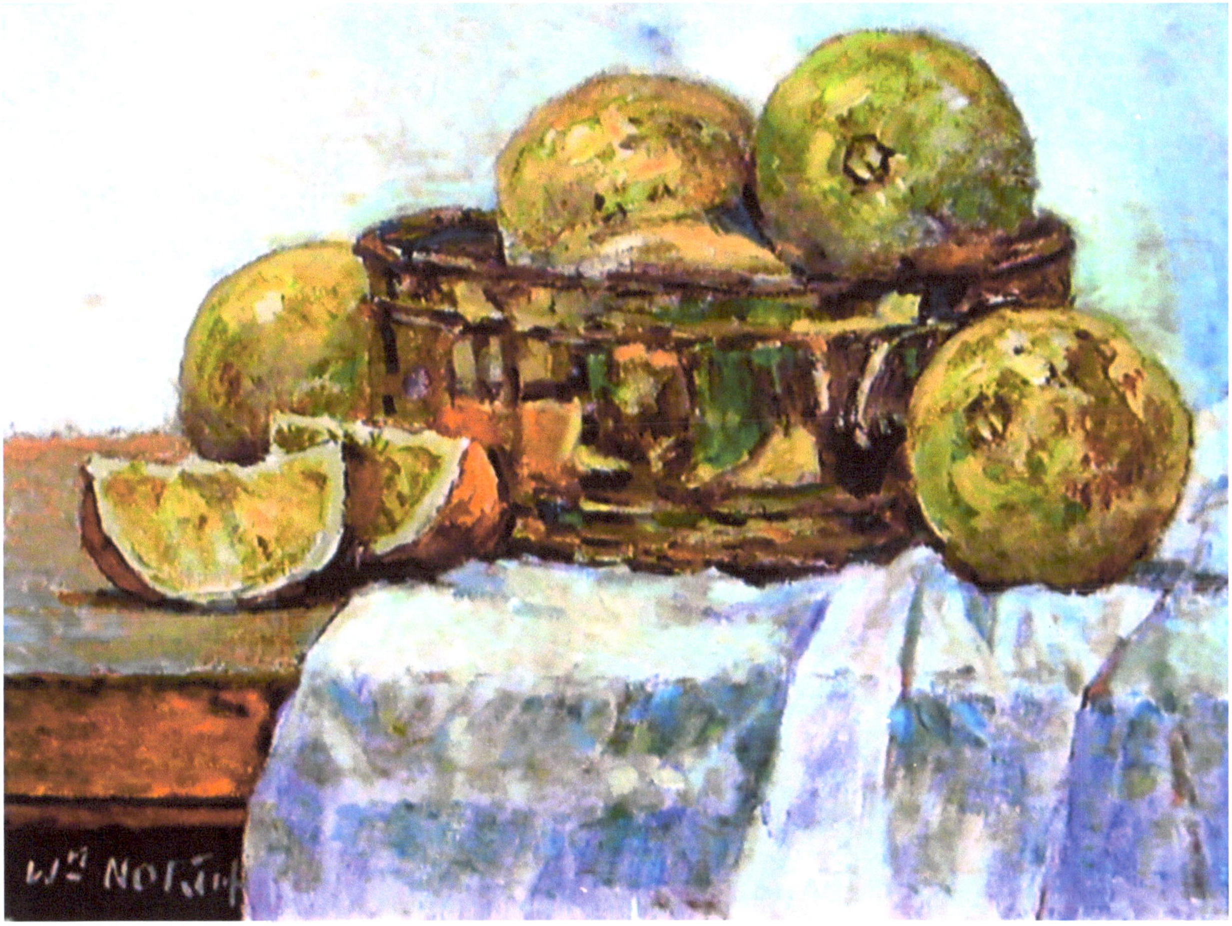

BOLD AND FRESH

Rhode Reds and Ruby Red

Originally, oranges were my still life subject of choice. They were a good scale for the size of my canvas as well as my collection of props and containers. Some of those reasons changed, along with my artistic viewpoint, when I was approached by two growers at the Citrus EXPO who asked why I never painted grapefruit. Once I explained my reasons, I promised to bring a grapefruit study to next year's EXPO. This painting is the result. I liked it so much that I put it into print, and it's a popular part of my permanent print inventory.

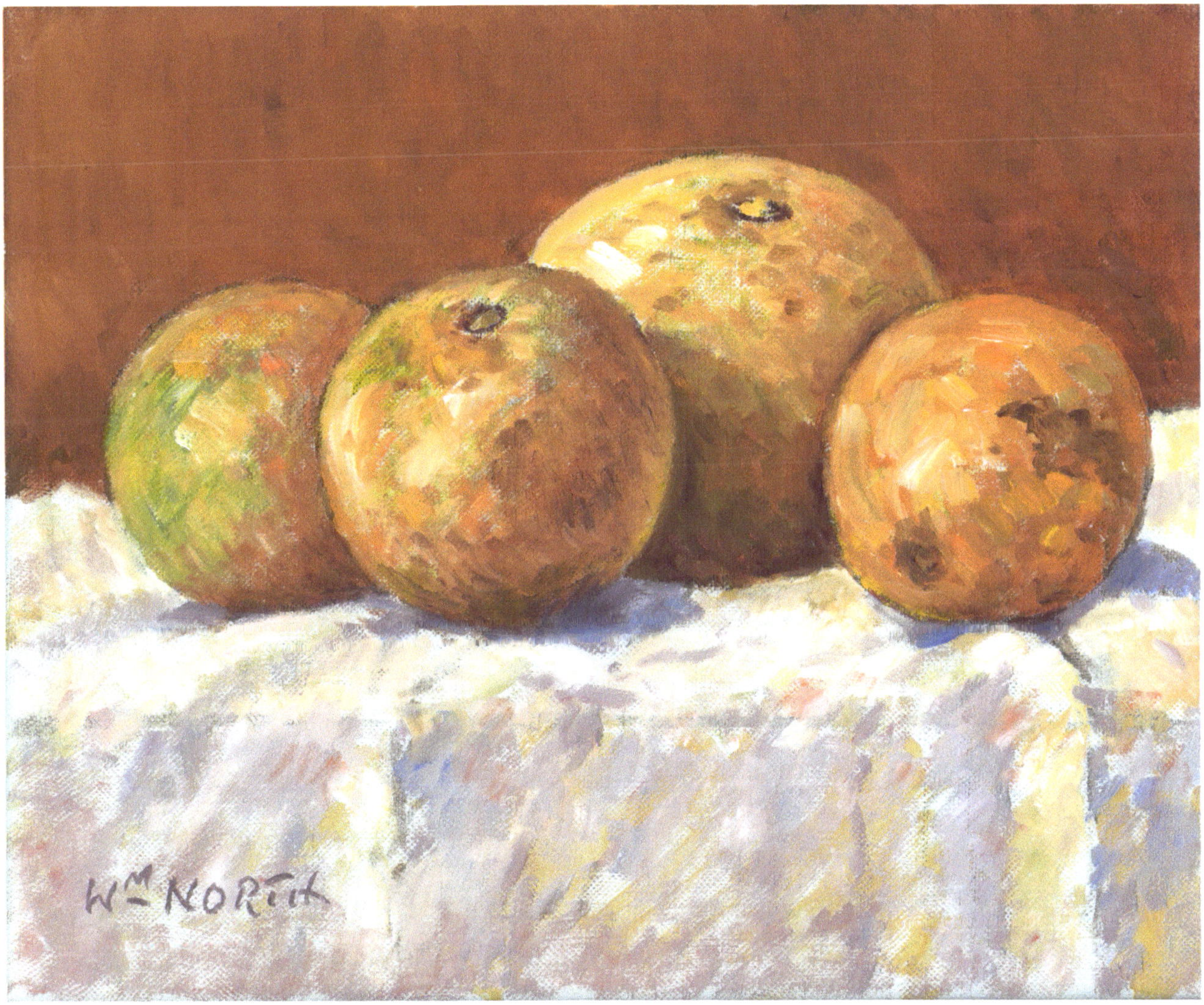

RHODE REDS AND RUBY RED

Florida's Natural

This is one of a kind; the only still life I ever painted showing juice cartons. Florida's Natural "Most Pulp" style juice was often consumed in our house.

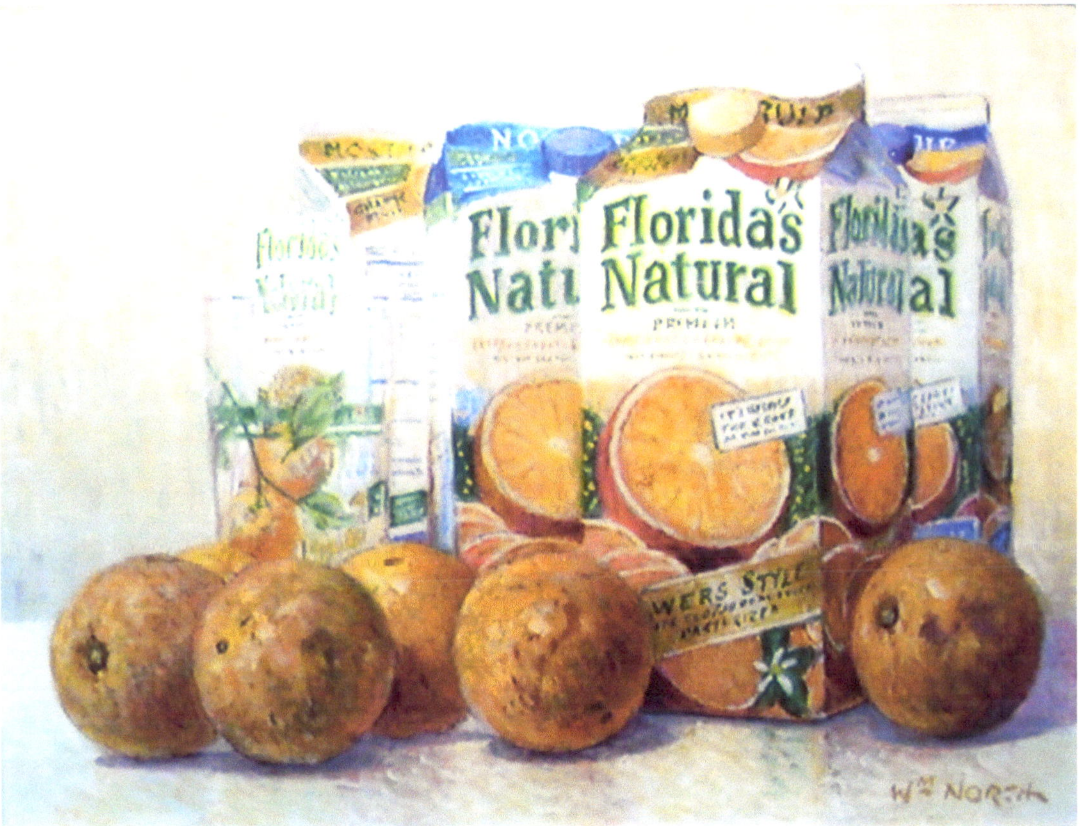

FLORIDA'S NATURAL

Battered and Bruised

The title refers to the often-used brass container, not the fruit. The oranges are moderately weathered, just enough to make them attractive to the artist's eye and to be included in this painting.

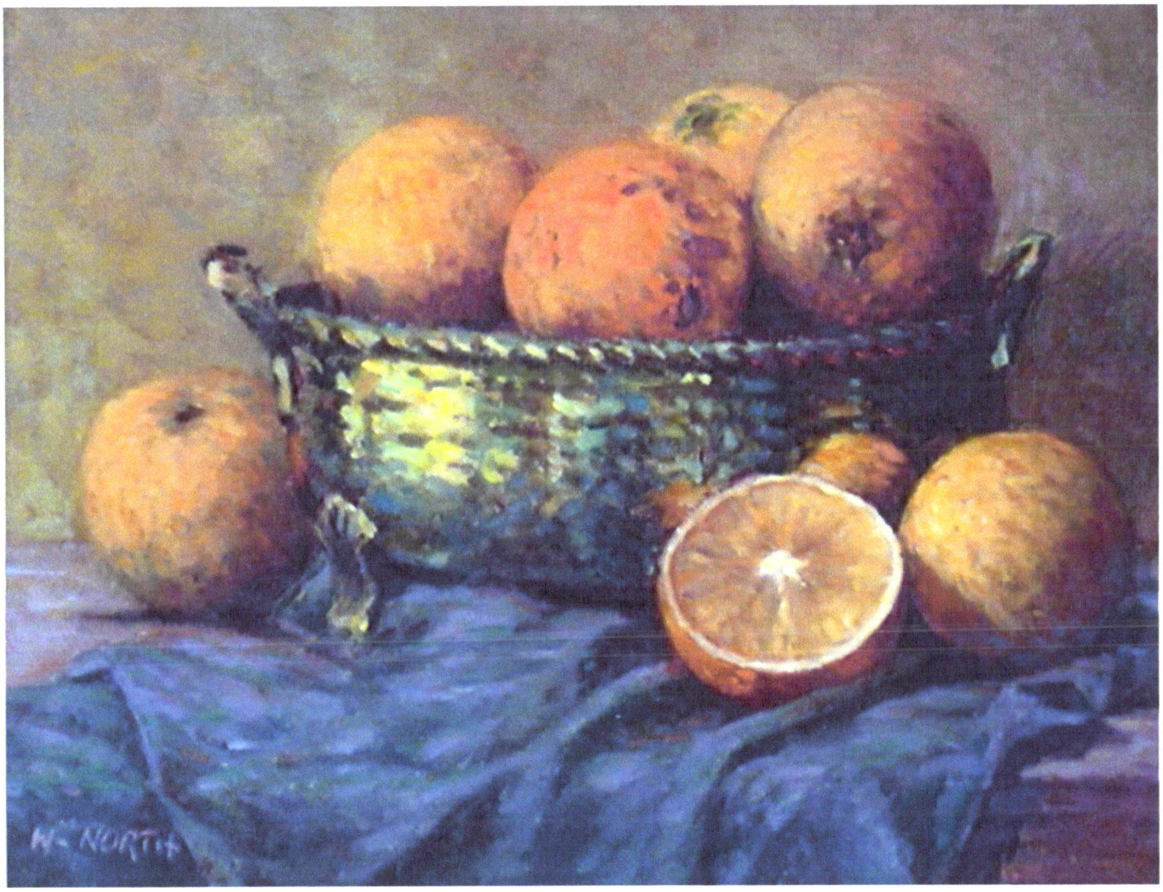

BATTERED AND BRUISED

The Other Brass Bowl

Neither hammered nor battered, this brass bowl came from my inventory of containers. It's size, shape, and handles always added zest to a painting. Real orange tree leaves are included in this arrangement as a tribute to the real tree.

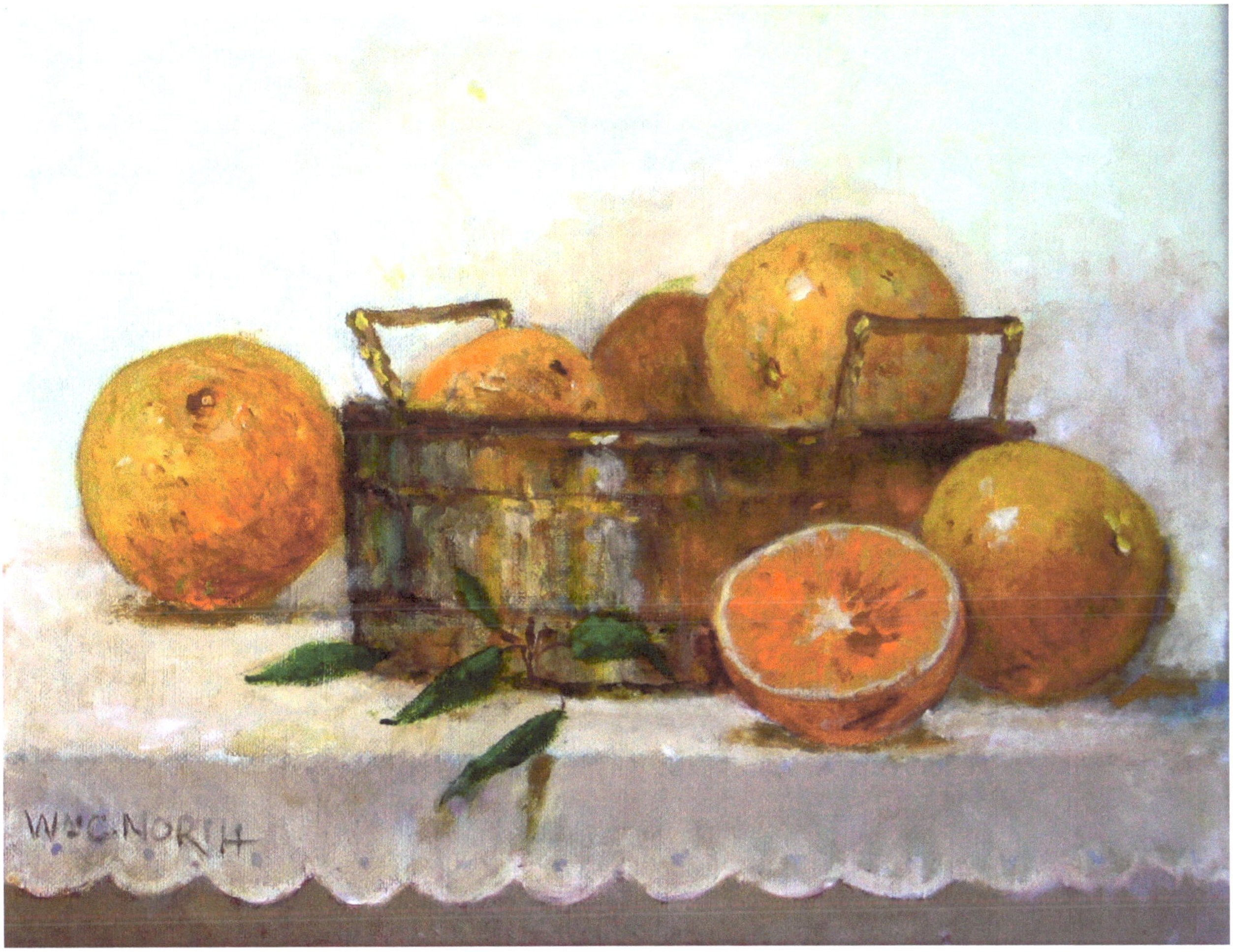

THE OTHER BRASS BOWL

Not Just Another Pretty Bowl

Sometimes the ripe orange was so brilliant that its container had to be modest. A straight-sided clear glass bowl served the composition well to show off all the oranges.

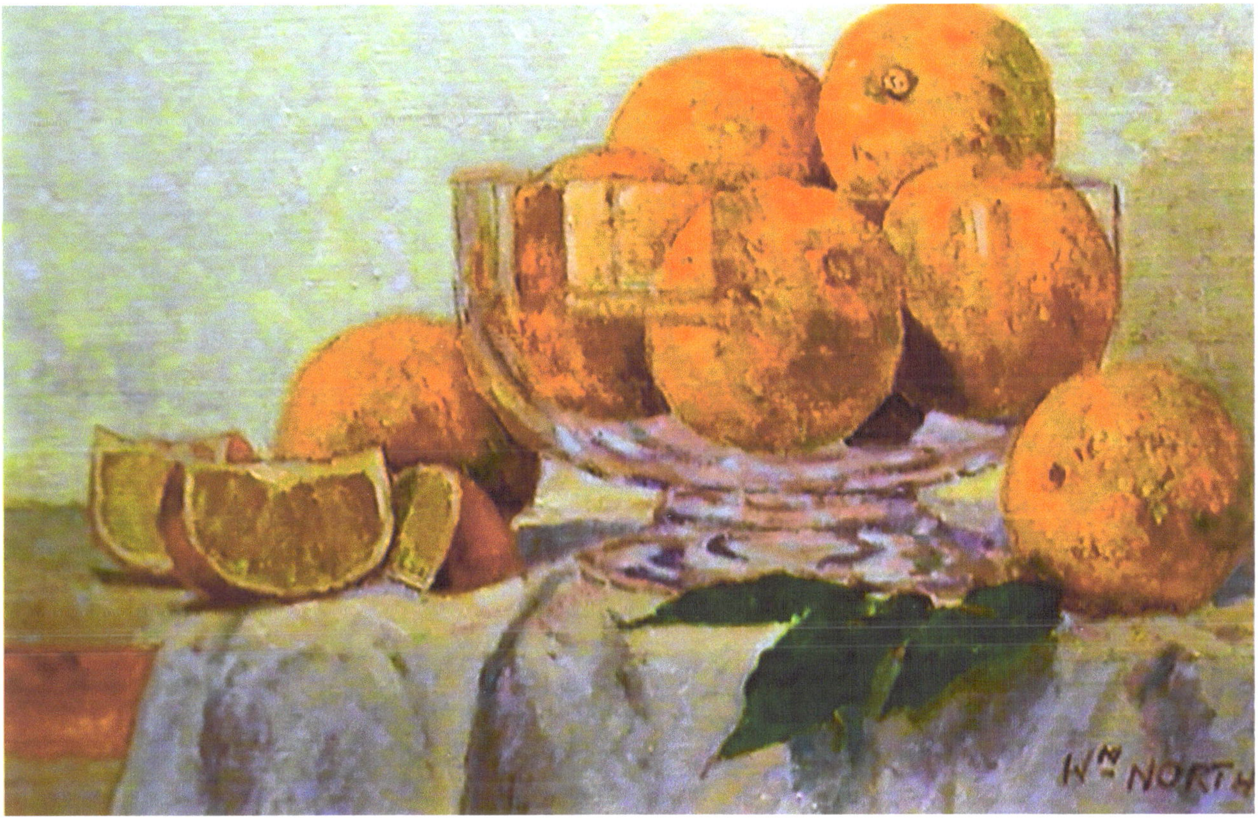

NOT JUST ANOTHER PRETTY BOWL

Soup Bowl Encore

The oriental soup bowl is back for an encore. Three wedges and three uncut oranges silhouette the background and foreground.

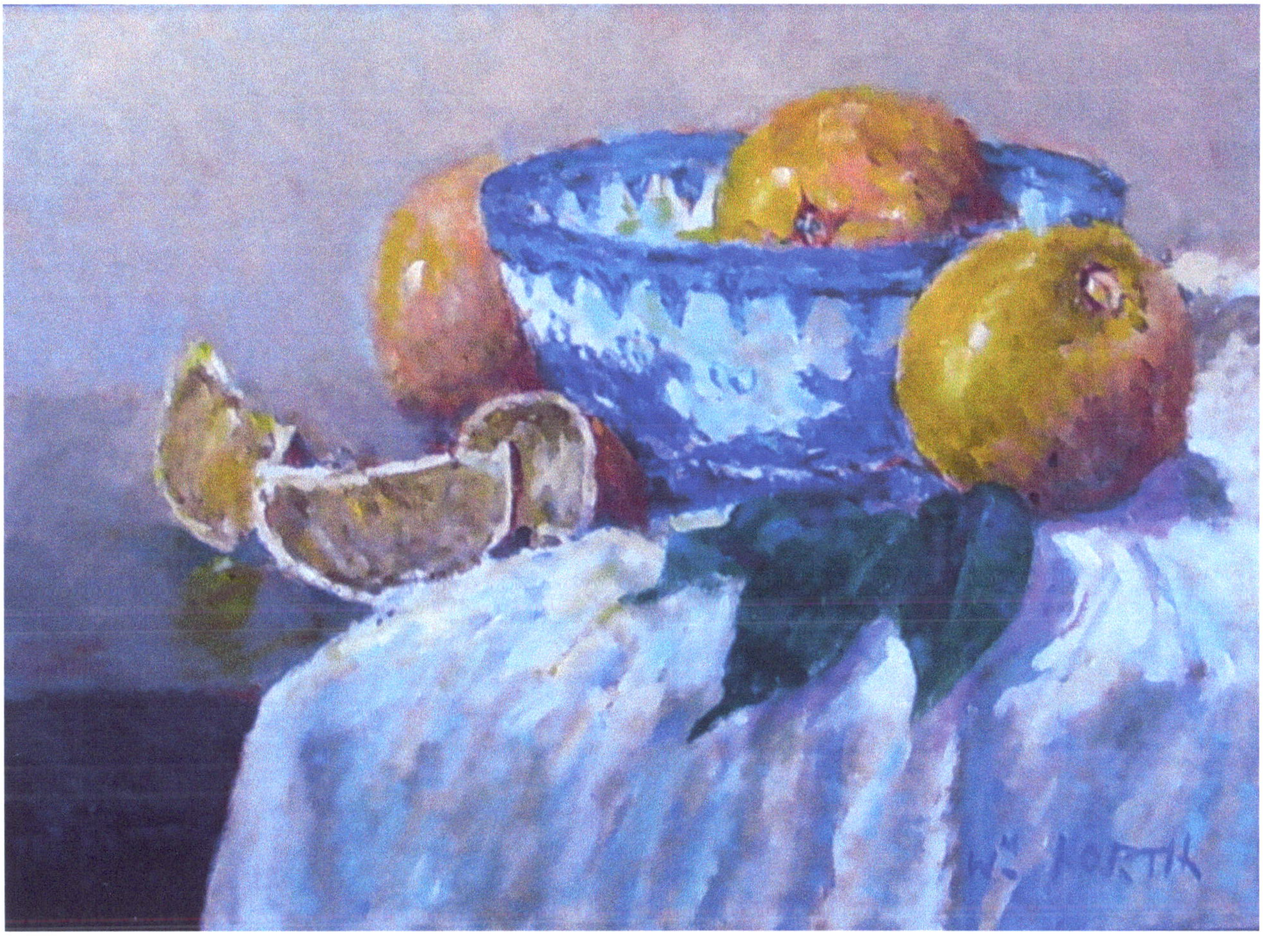

SOUP BOWL ENCORE

Oranges at Length

Oranges, oranges everywhere in this stretch version of my standard still life arrangement. The 10″ x 20″ dimension makes it unique in my portfolio, and its unusual length was my way of trying something new.

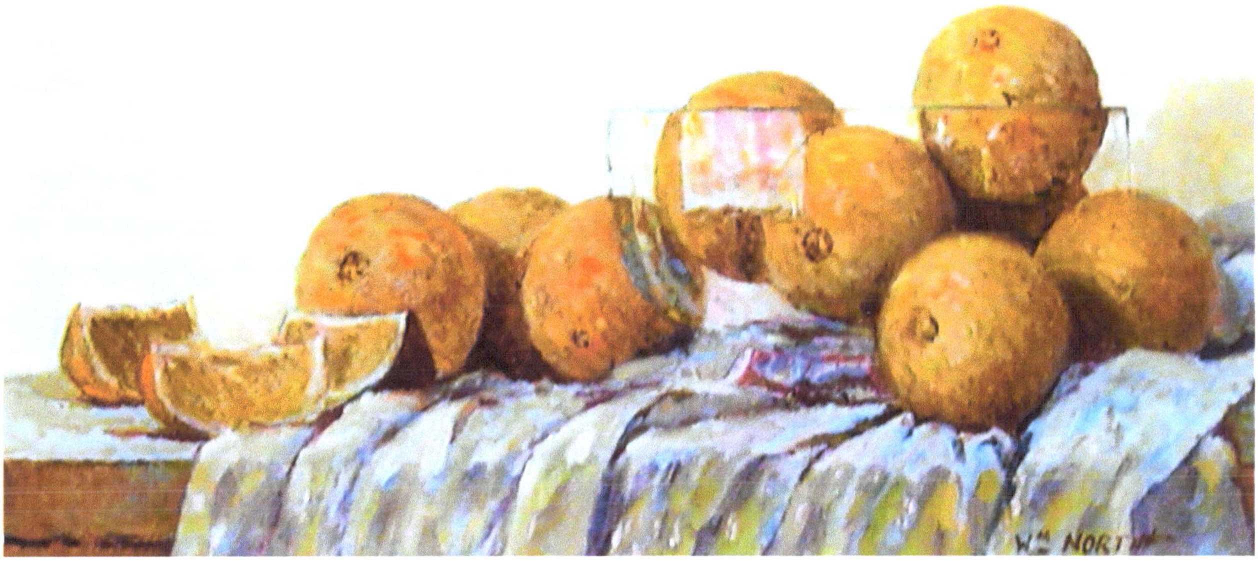

ORANGES AT LENGTH

The Brown Paper Bag

The man at the Cabbage Patch produce stand handed me the oranges I'd selected in a brown paper bag. My creative juices were sparked at the sight of this sturdy paper container, and I knew I'd use it in a painting some day. After completing many compositions using glass, brass, and wooden bowls, the bag as a prop provided a restful change.

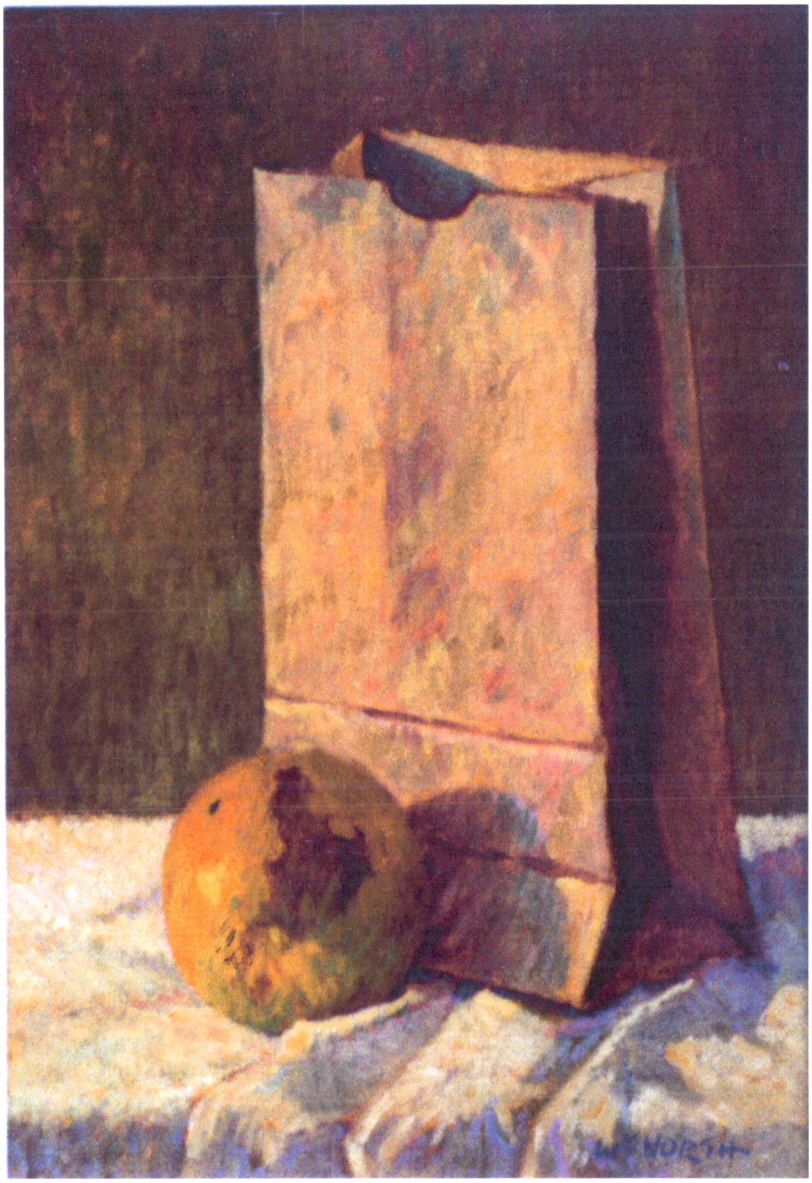

THE BROWN PAPER BAG

The Full Monty

This mature work shows it all—a nice grouping of battle-scarred fruit, excellent treatment of the distortions in the fluted bowl, a nicely draped cloth, and extra detail in the wooden tabletop.

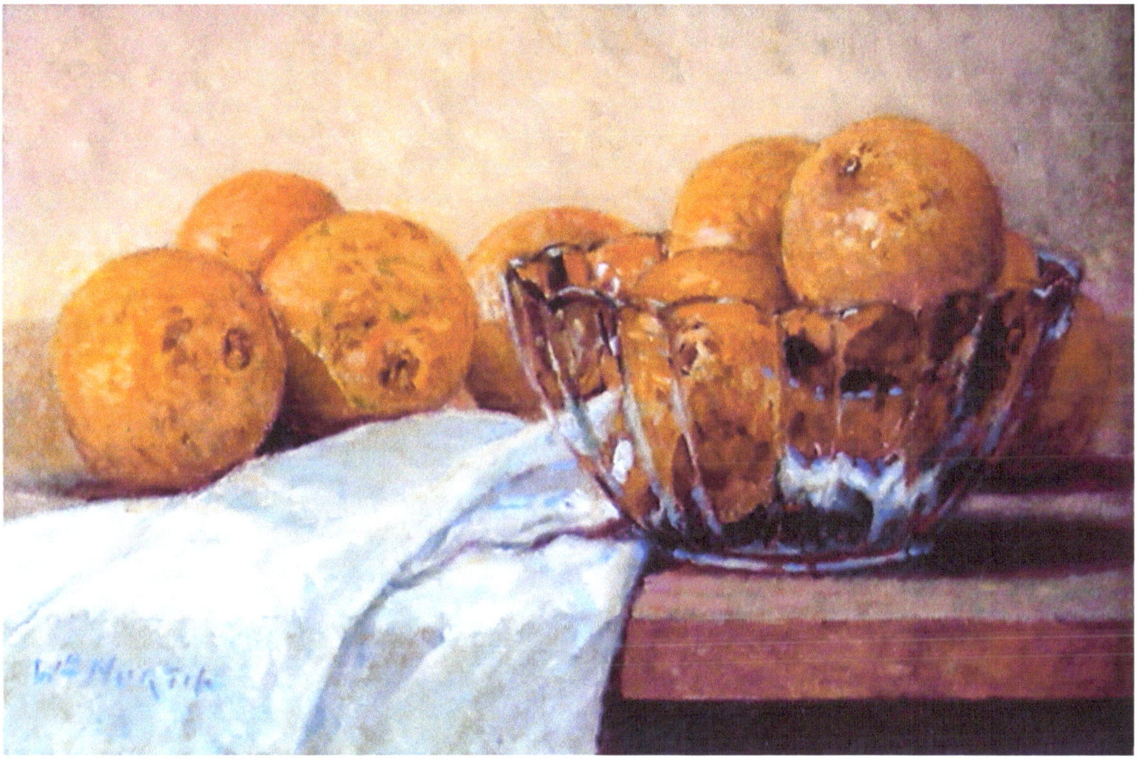

THE FULL MONTY

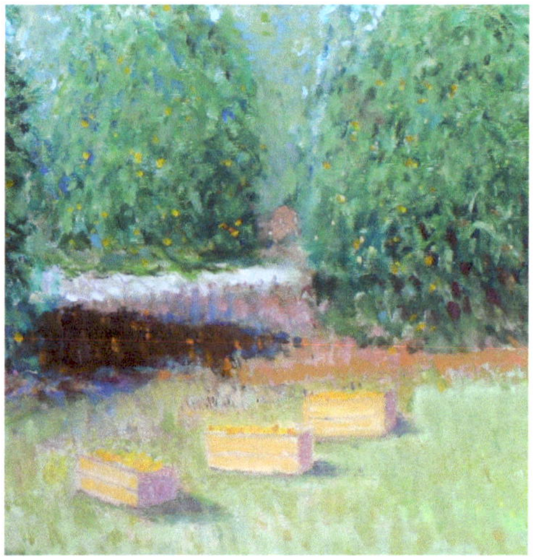

Orange Groves 5

Painting the Land of the Golden Harvest

Painting in Paradise

The start of another great day in paradise.

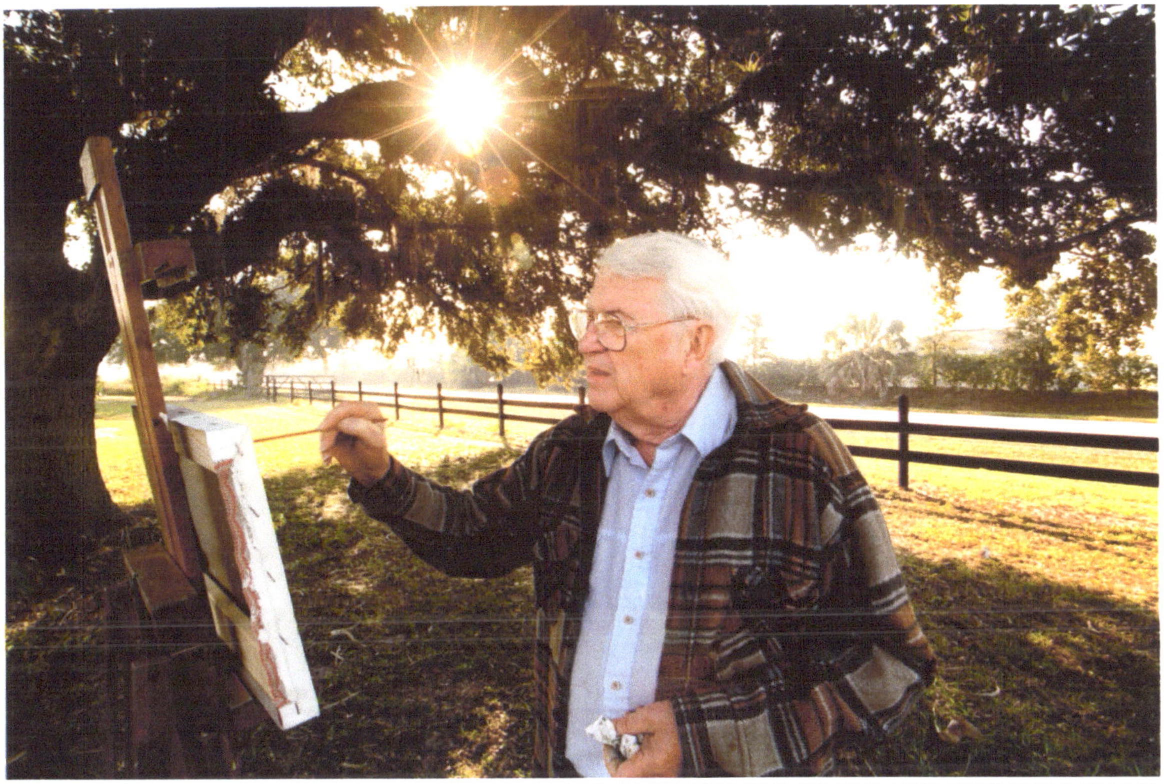

PHOTO COURTESY OF NAPLES DAILY NEWS/THE BANNER

Orange Grove with Boxes

This orange grove scene holds a special place in my heart. It is one of my first grove scenes, and it is the first with wooden crates. The trees were painted first, then I decided to add the crates. I placed them every which way on the ground, creating a complex perspective problem. Anyone can tell if the perspective is not just right. With no drawing, just brushwork, I put in the boxes perfectly. This painting had a happy ending, which makes it one of my favorites.

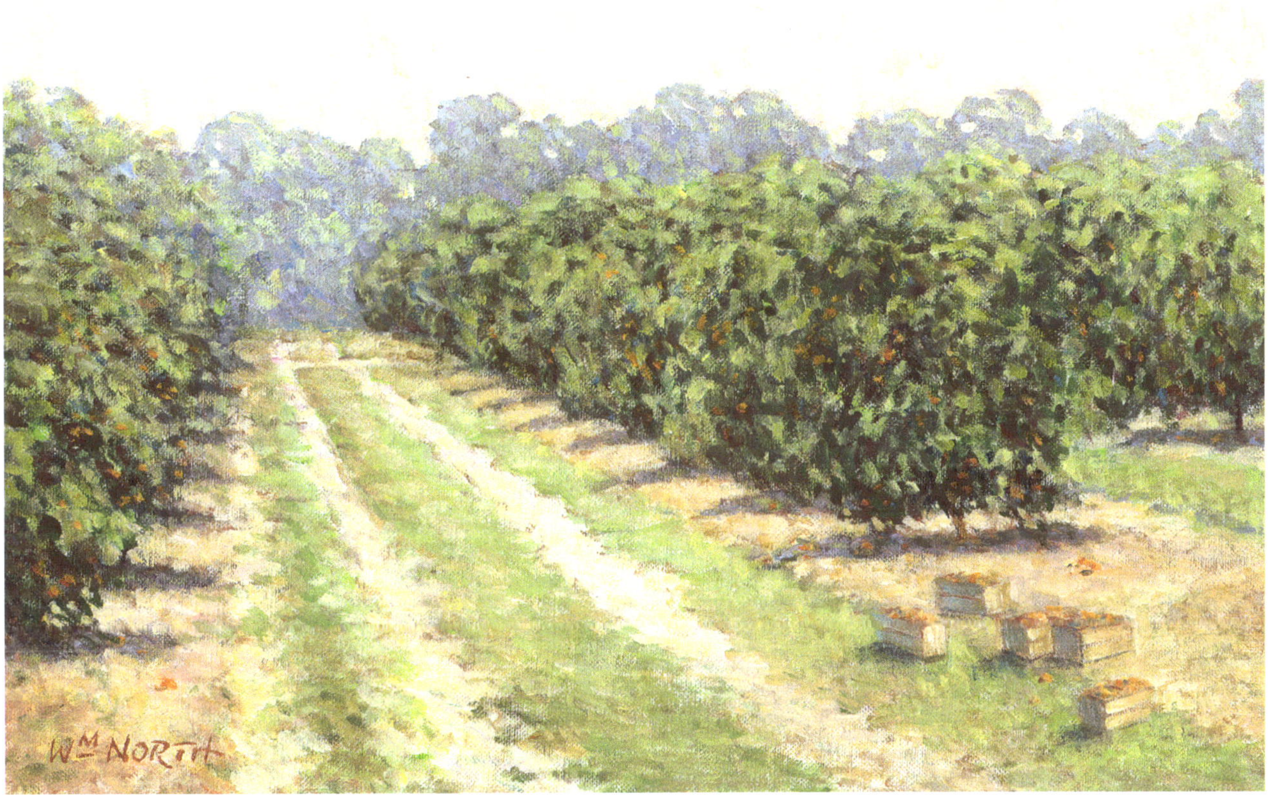

ORANGE GROVE WITH BOXES

My Shade Tree Friend

Call me a tree hugger.

I became close friends with this tree in the Spiegel Grove since it provided a shady spot where I could set up my easel in comfort. I painted many different views from under this tree and then decided it deserved its own portrait—even if I had to stand in the blazing Florida sun to accomplish it.

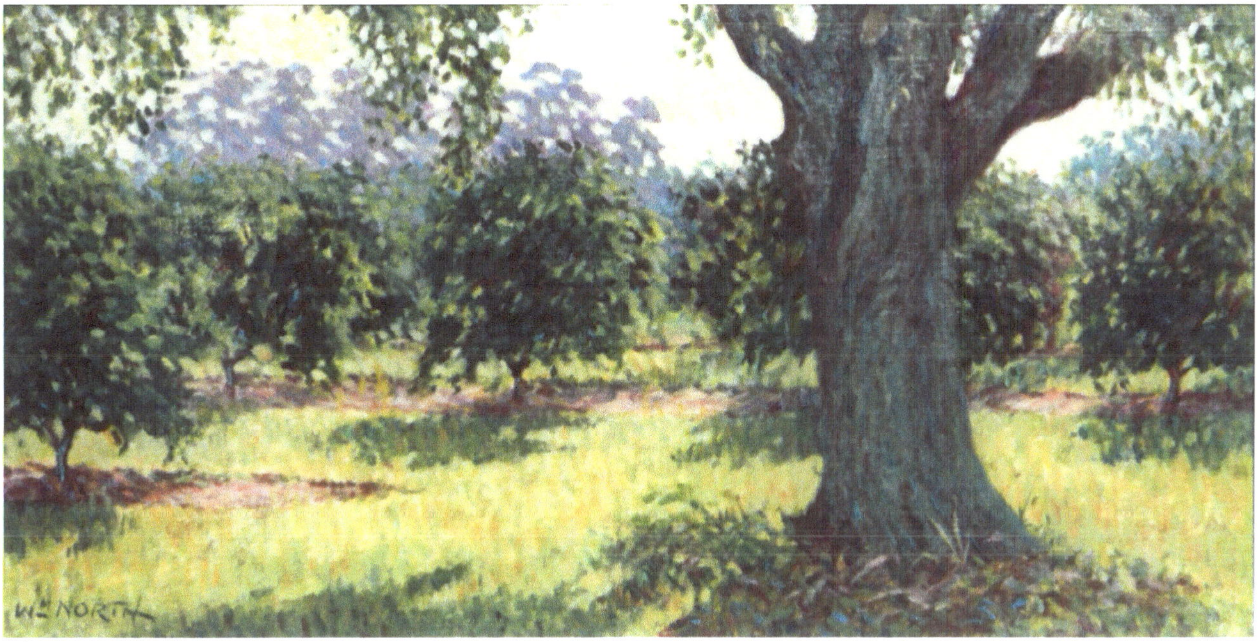

MY SHADE TREE FRIEND

Harvest Time

Our picker, standing on a ladder, is working in a small grove the old-fashioned way, picking one orange at a time. Most Florida groves still harvest this way: with an army of pickers in the trees.

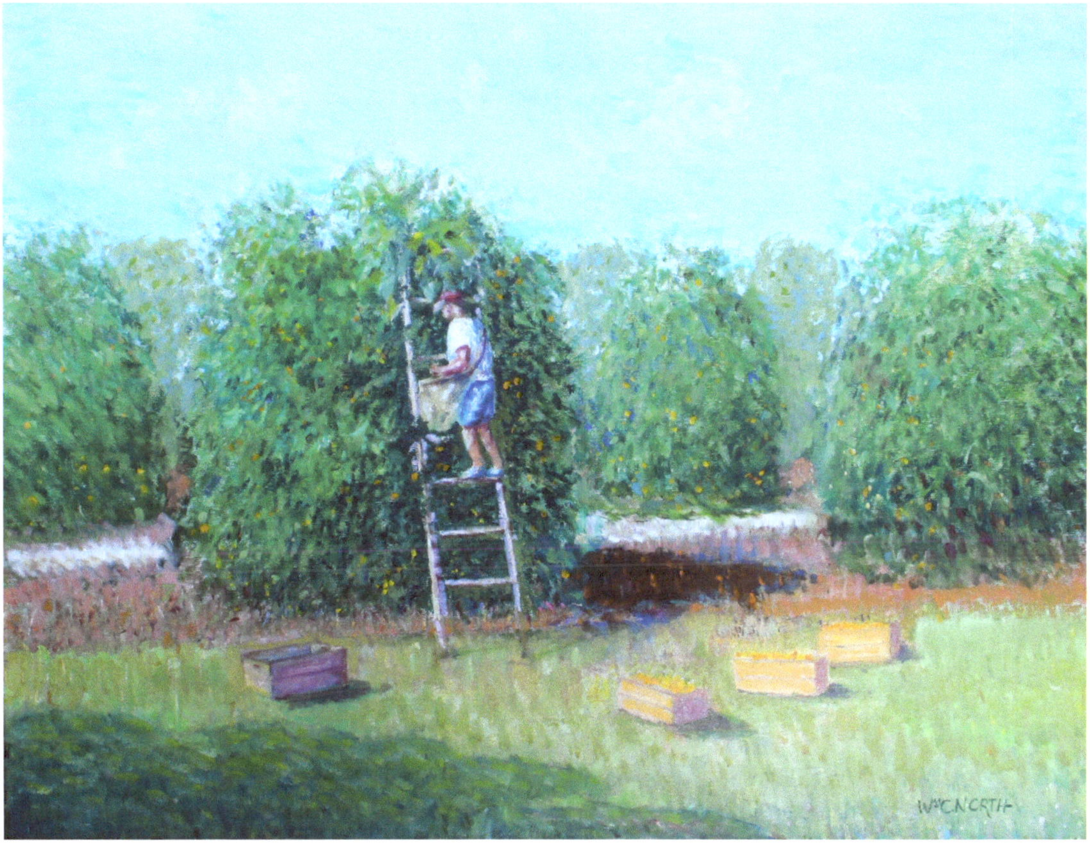

HARVEST TIME

Red Pickup Truck

The red pickup parked in a grove is a new subject in my work. Handling the truck was a challenge with my limited vision, especially the perspective. In my better days, I could have drawn it with ease. It took four tries to get it right; I even consulted with members of my family. Some said the tailgate was too wide, and some said too narrow. There was a dispute about the wheels being off the track. It finally came together.

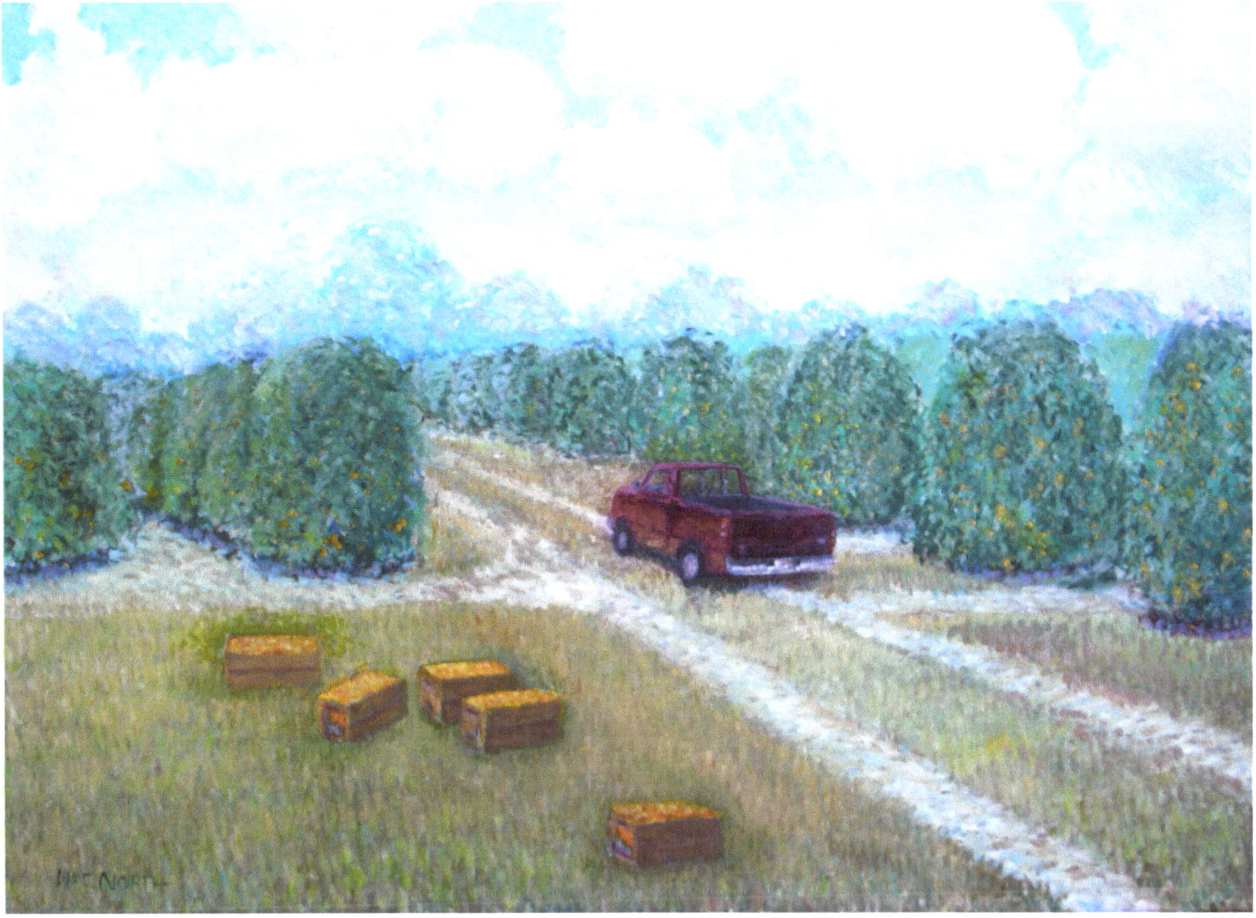

RED PICKUP TRUCK

A Tree For All Seasons

Each orange variety has its season.

Only the tree knows the reason.

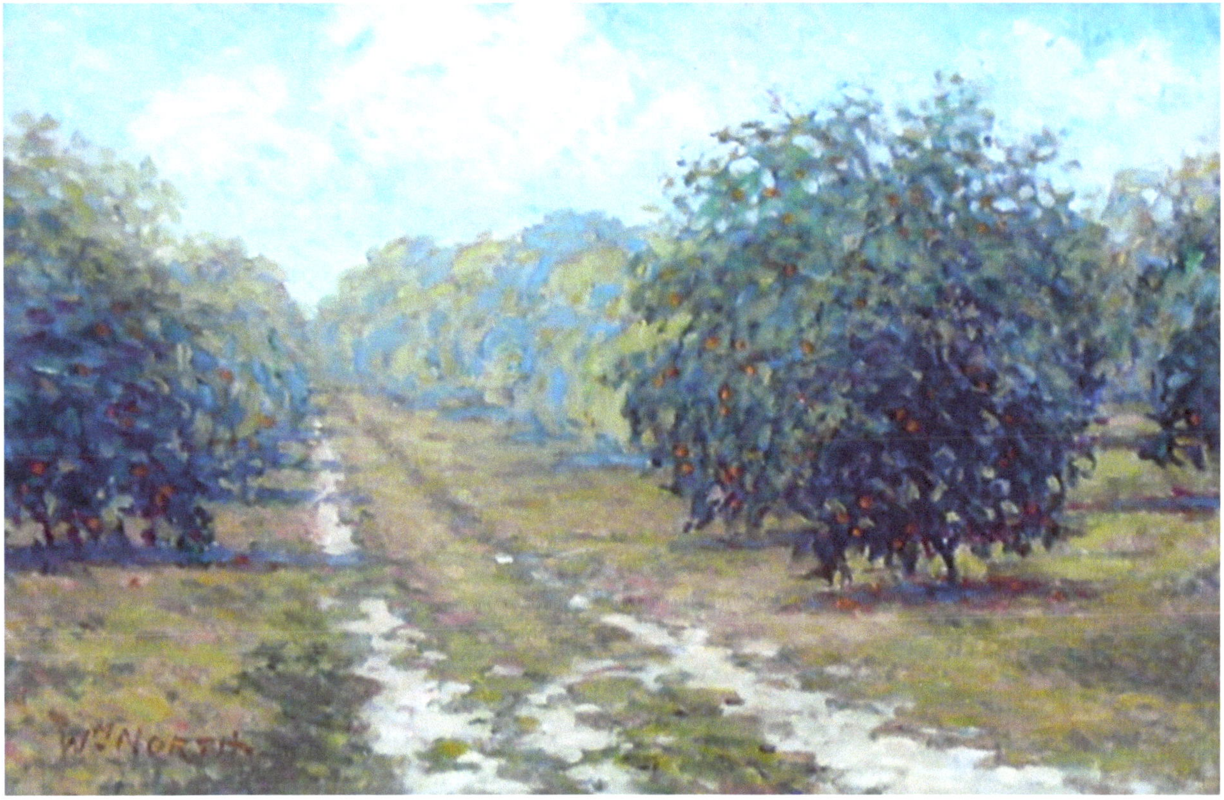

A TREE FOR ALL SEASONS

In the Eye of the Beholder

These well-cared-for trees shimmer in the Florida sun. When I planned the painting, I wanted to bring out all the inherent grace and beauty I saw there. I used soft edges and a lightly toned palette. I feel it conveys the warm, subtle feeling I had while painting it.

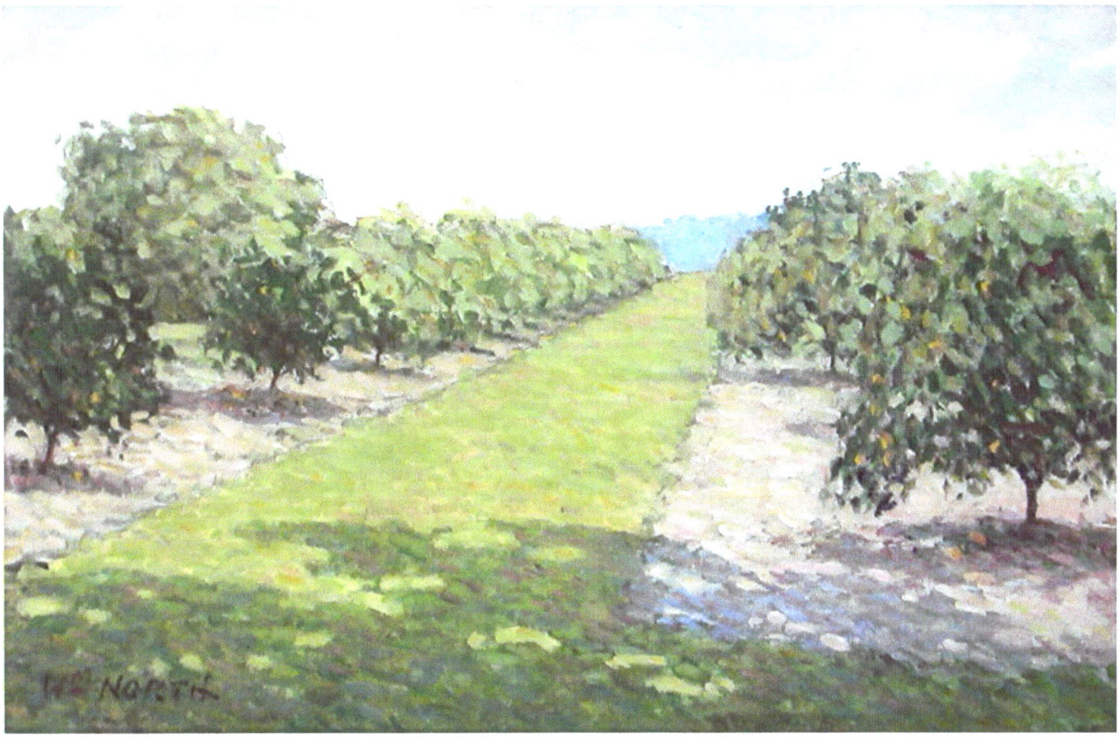

IN THE EYE OF THE BEHOLDER

The Alva Grove

The tree line silhouette in the background identifies this as George Austin's grove in Alva. I painted it precisely as it appeared that day. Other groves have similar tree lines, but each has its distinctive outline.

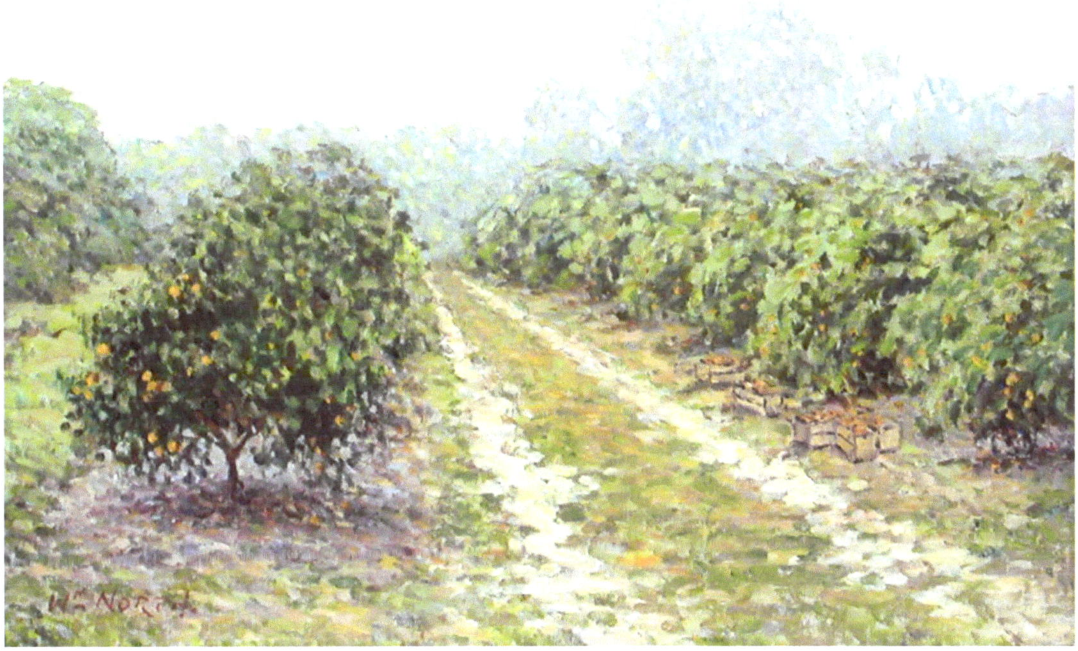

THE ALVA GROVE

Shadows

Shadows always tell a story. They tell the approximate time of day, clarity of the air, and the general east-west bearings. This scene captures the grove in early morning at sunrise.

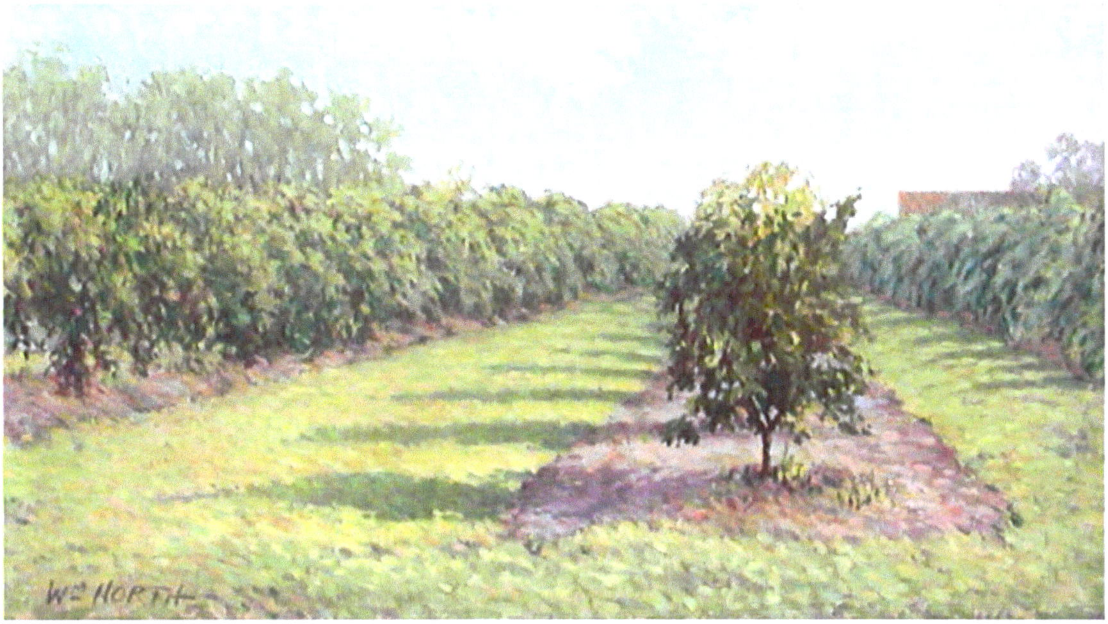

SHADOWS

Truck at the End of the Road

The 1937 Ford truck is the focal point of this painting. I did not draw it first but took a brush to paint the headlights and hit just the right spots. With a few strokes, I created an impression of the truck, which seems to show the right amount of detail.

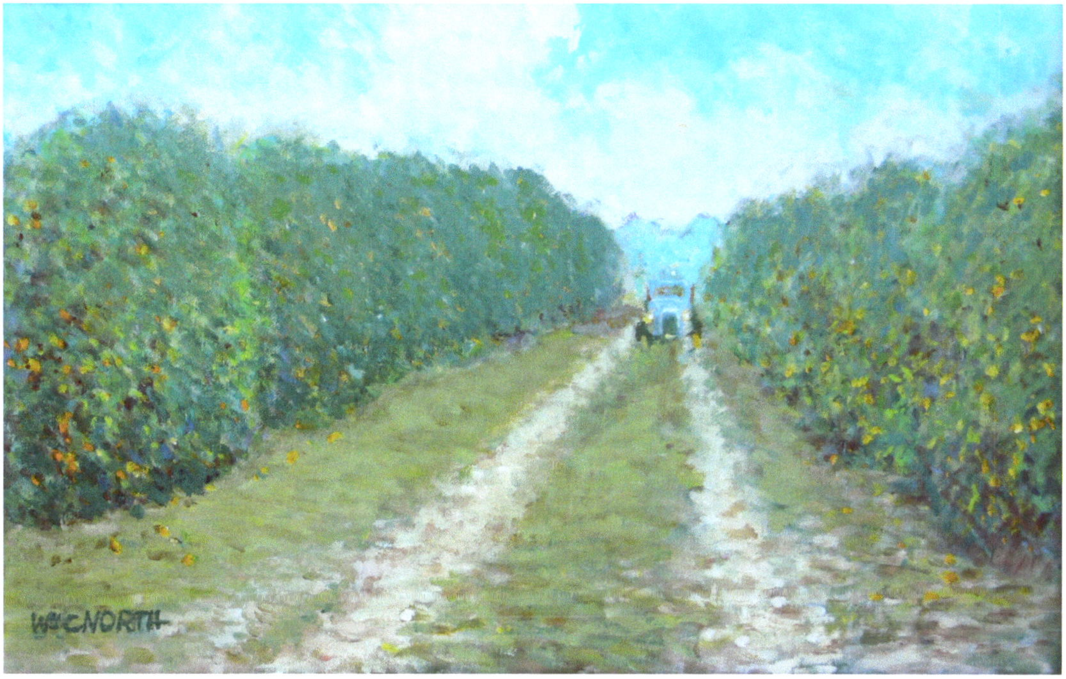

TRUCK AT THE END OF THE ROAD

Morning Shadows

The early morning light and the shadows cast by the rising sun make this corner of the grove a place of beauty.

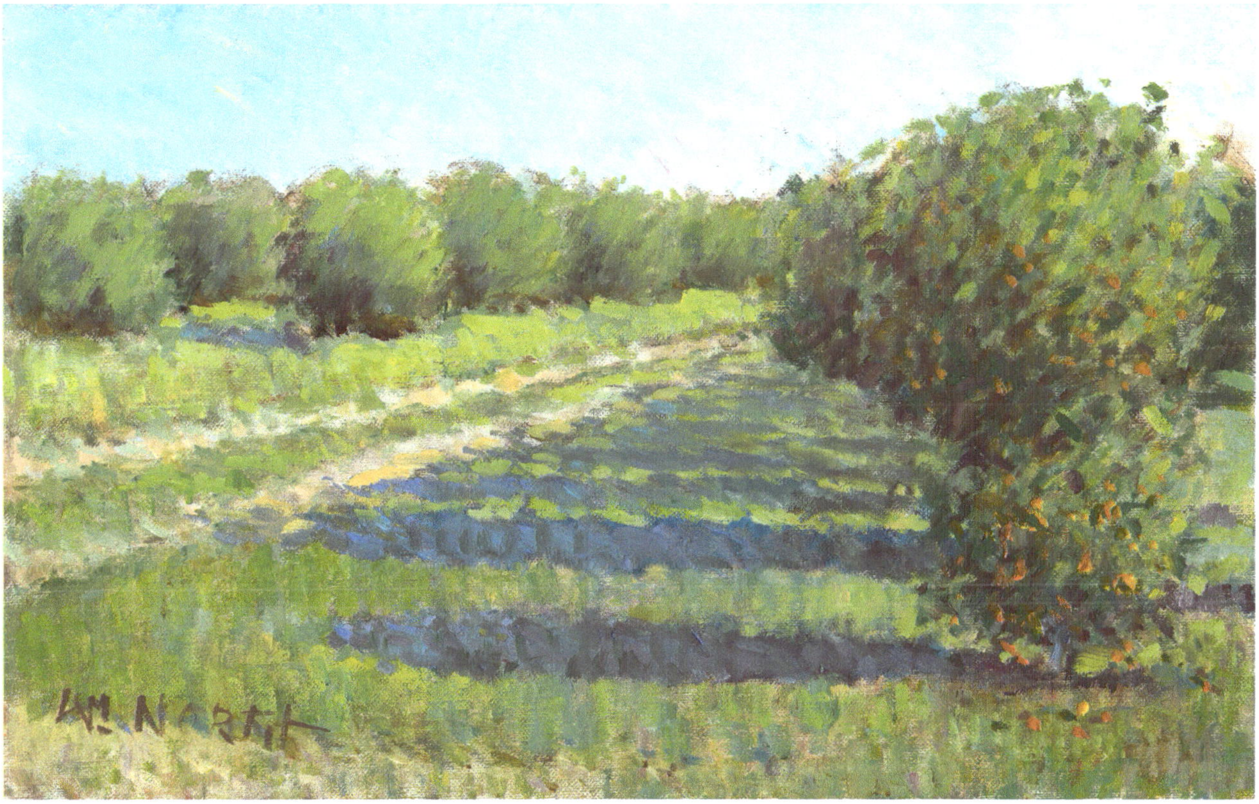

MORNING SHADOWS

Florida Citrus Grove

The Gulf Citrus Growers Association chose this painting as a 25th Anniversary Commemorative Print (1986–2011). The painting is titled "Florida Citrus Grove." Reproduction prints were made of the original oil, which I signed above the line "From an Original Oil Painting by William North," before they were framed and presented at the award ceremony.

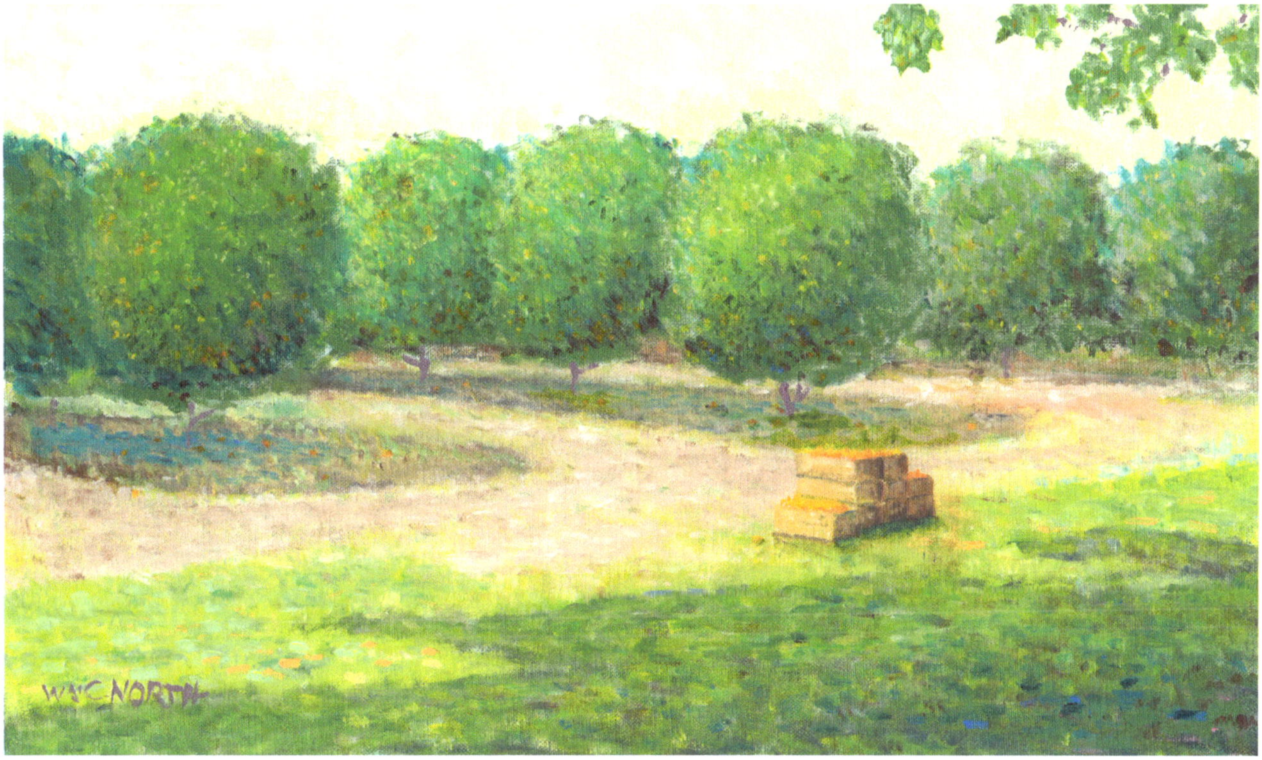

FLORIDA CITRUS GROVE

Name That Orange

I did not record the variety of orange in this painting. They might be Valencias, Navels, Temples, Hamlins, Arcadias, Blood Oranges, Weldons, Bessies, Centennials, Dillars, Drake Stars, Early Oblongs, Boones, Enterprises, Jaffas, Golden Buckeye, Golden Nuggets, Parson Browns, Majorcas, Osceolas, Washington Navels, Magnum Bonums, The Robertsons, Starks, Pope Summer Nonpariels, Fosters, Rubies, Whittakers, Thomson Rhode Reds, Sanford Bloods, Homosassas, Whites, The Surprises, or something else. You pick it.

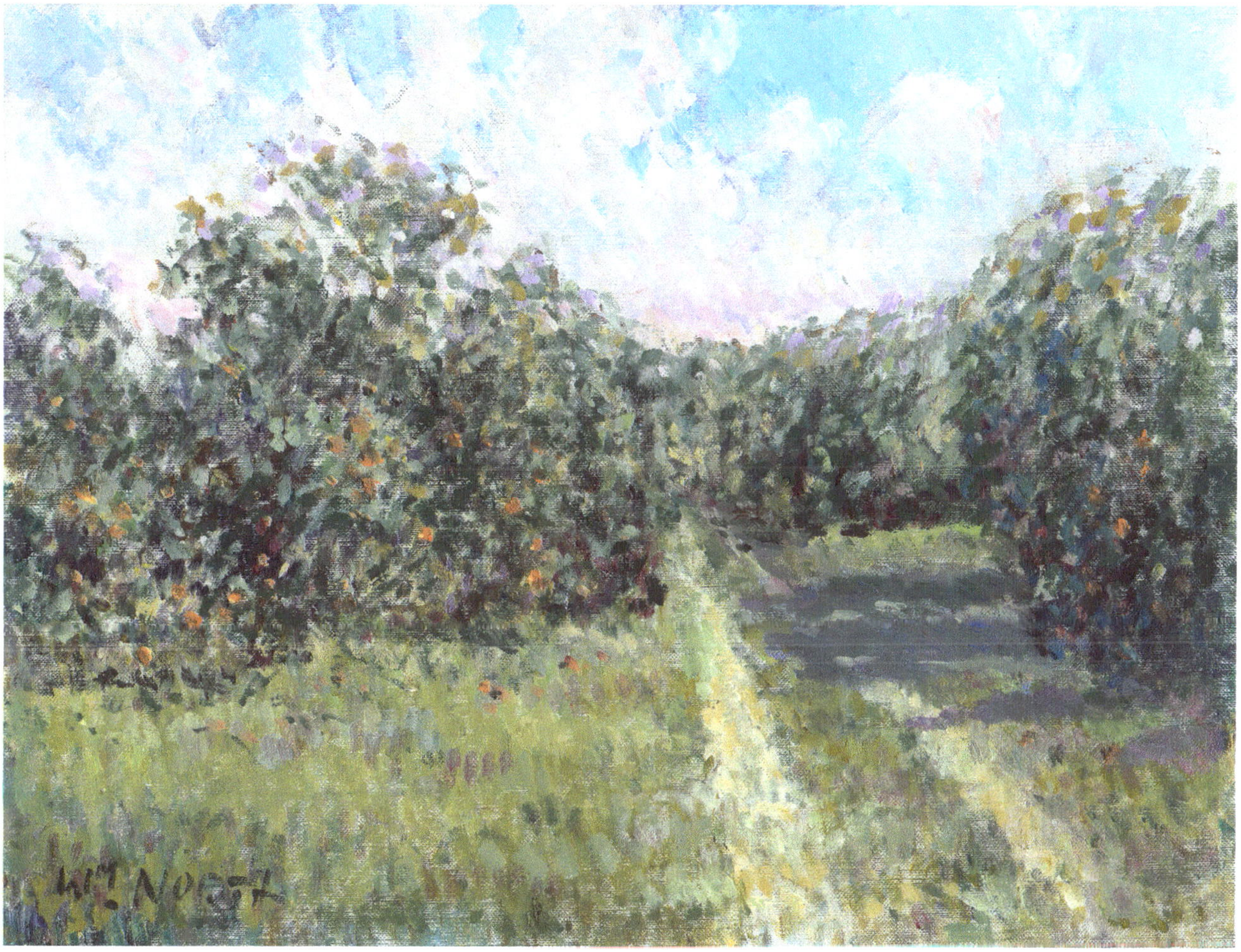

NAME THAT ORANGE

New World Oranges

When Christopher Columbus discovered the New World, the new world discovered oranges. There was no citrus here until Columbus brought the seeds on his voyage of discovery. Later, the Spanish spread the seed throughout their areas of exploration.

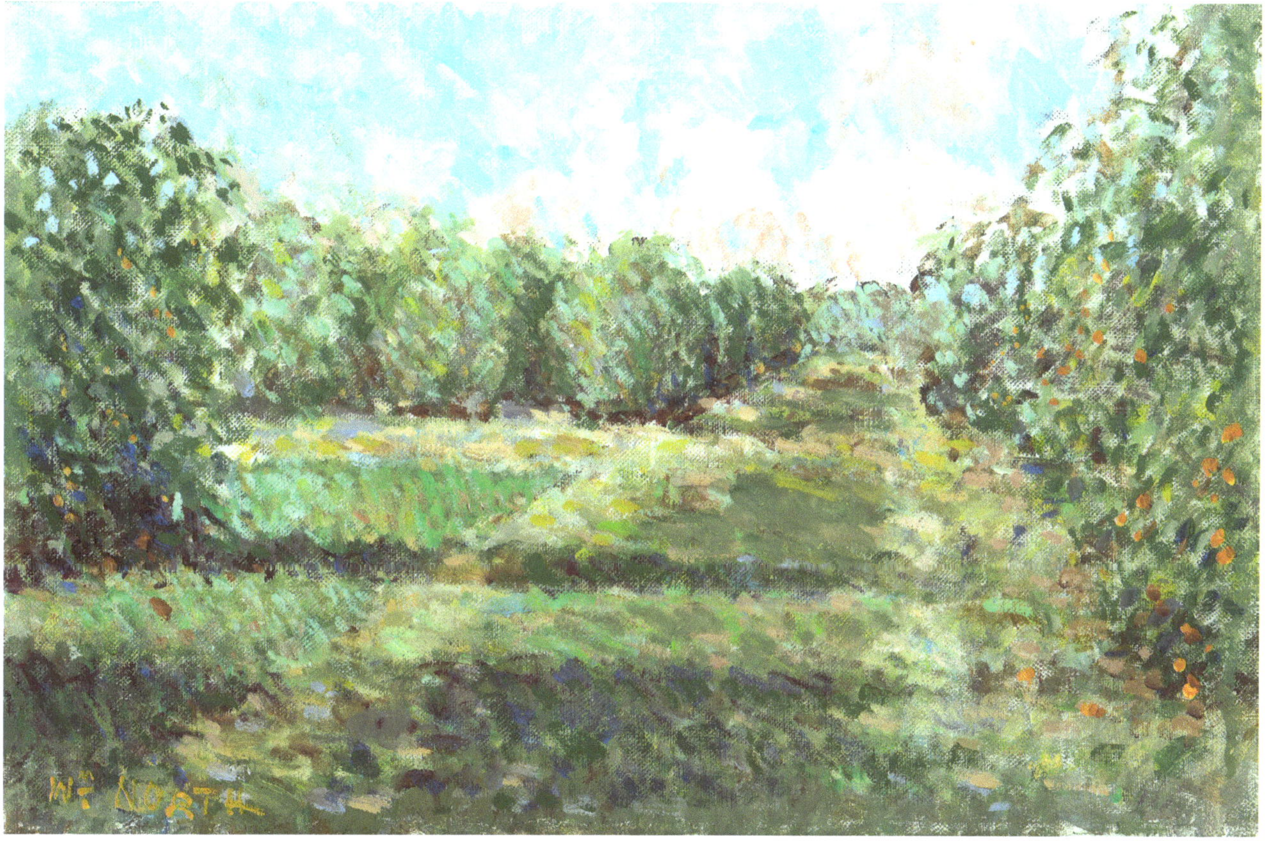

NEW WORLD ORANGES

True Orange

When seen from a distance, oranges appear as faint dots among a sea of green. So my challenge was to handle them in a way that looked natural and appealing. In my earliest grove landscapes, I didn't even paint them in.

As I became more confident in the subject, I found ways to add those tiny dots of orange that ring true and help to tell us about the season.

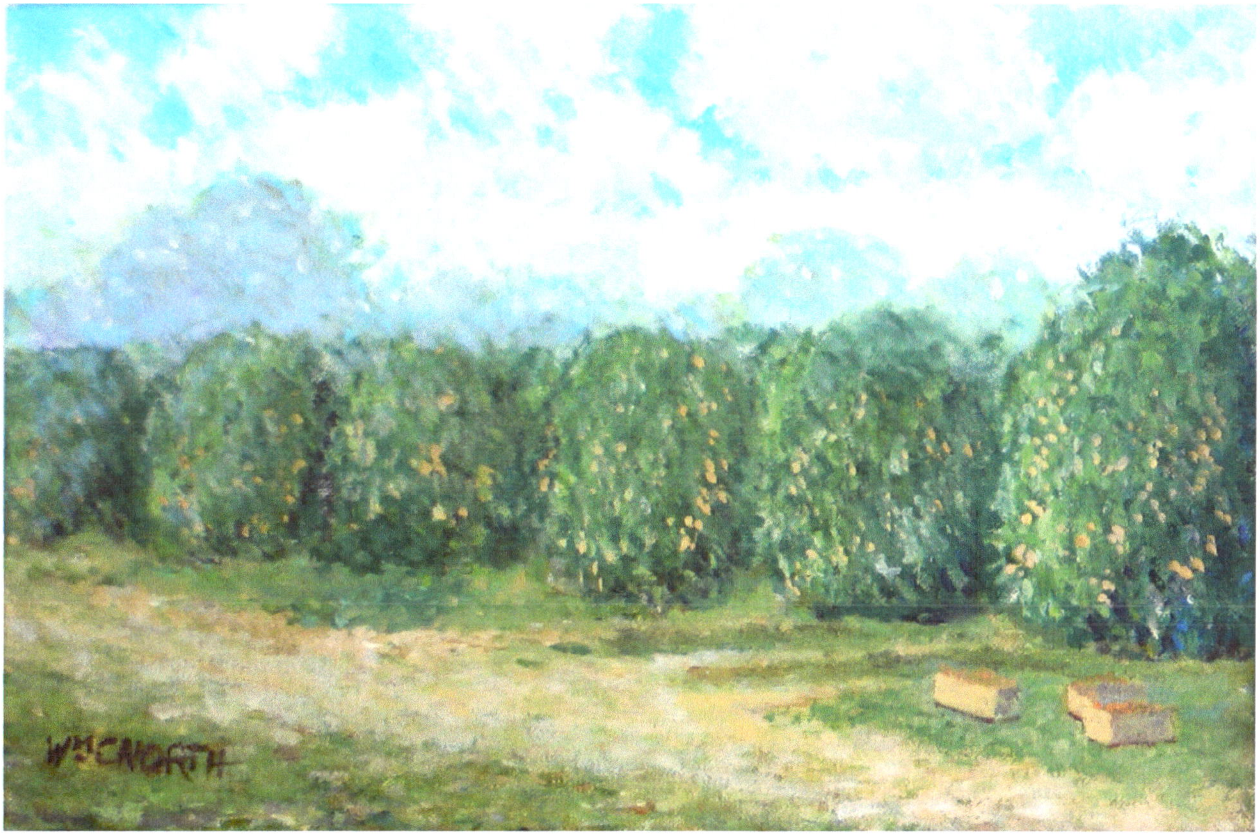

TRUE ORANGE

Florida Cattle Ranch

Folks from out-of-state don't know that Florida is a major producer of beef cattle and that there are large spreads here. Some ranchers also grow citrus, which is why I included this painting in the book. I painted this after I was declared legally blind, and so it was completed entirely in the studio.

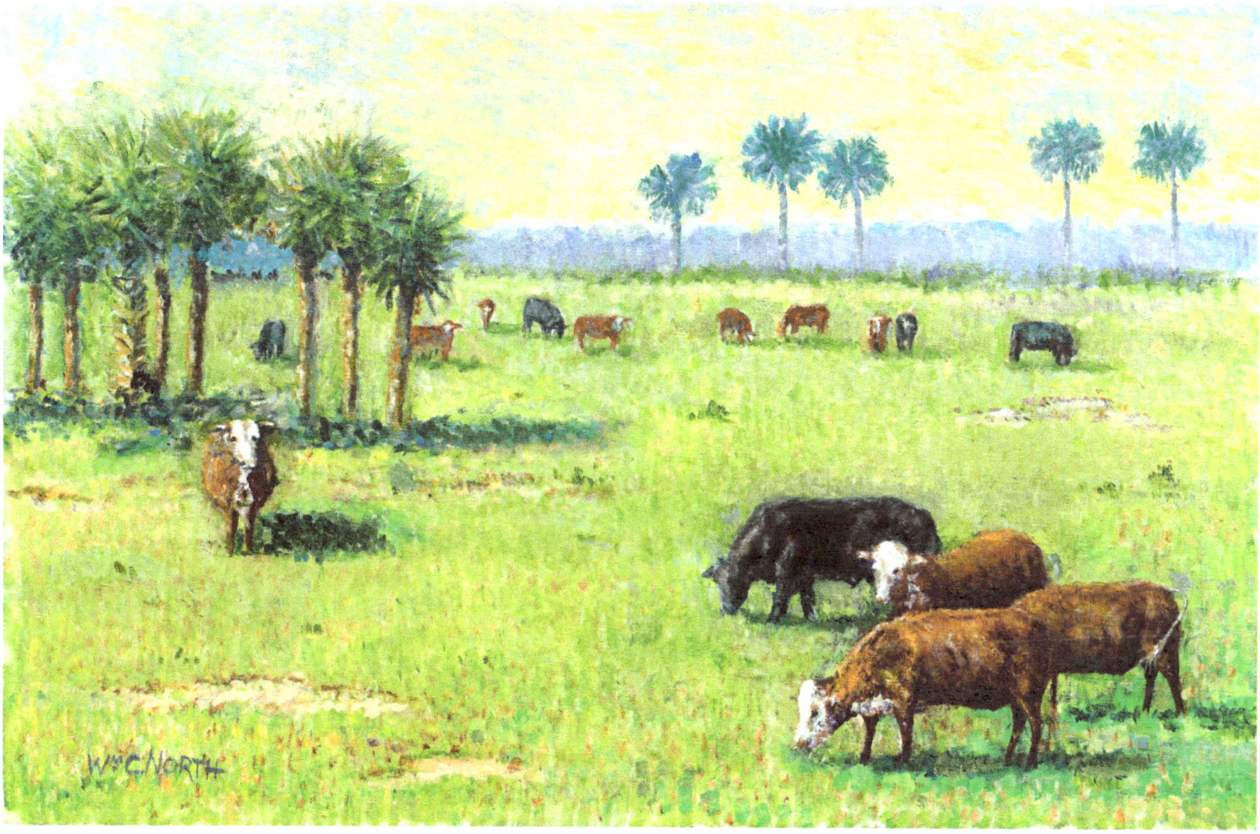

FLORIDA CATTLE RANCH

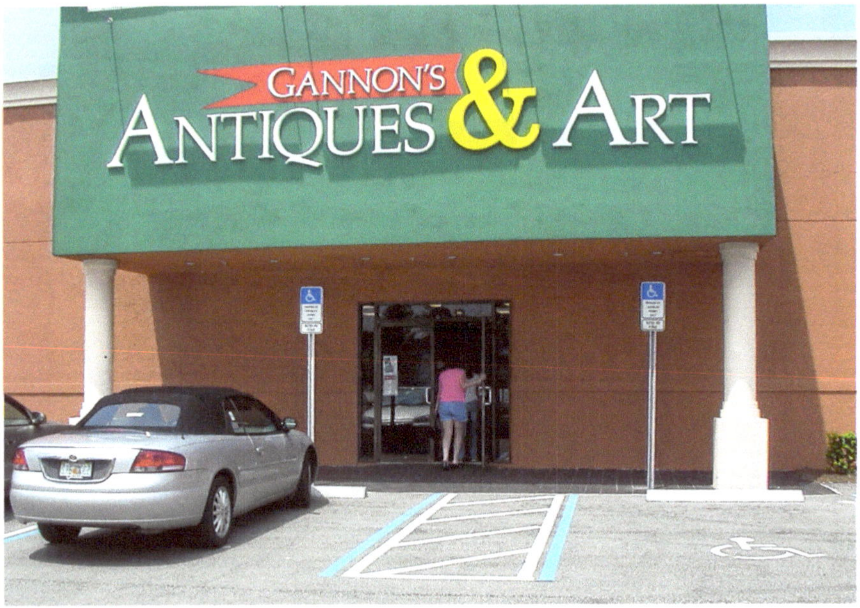
GANNON'S
ANTIQUES & ART

Wm. North Gallery

6

The Home of Florida Citrus Art

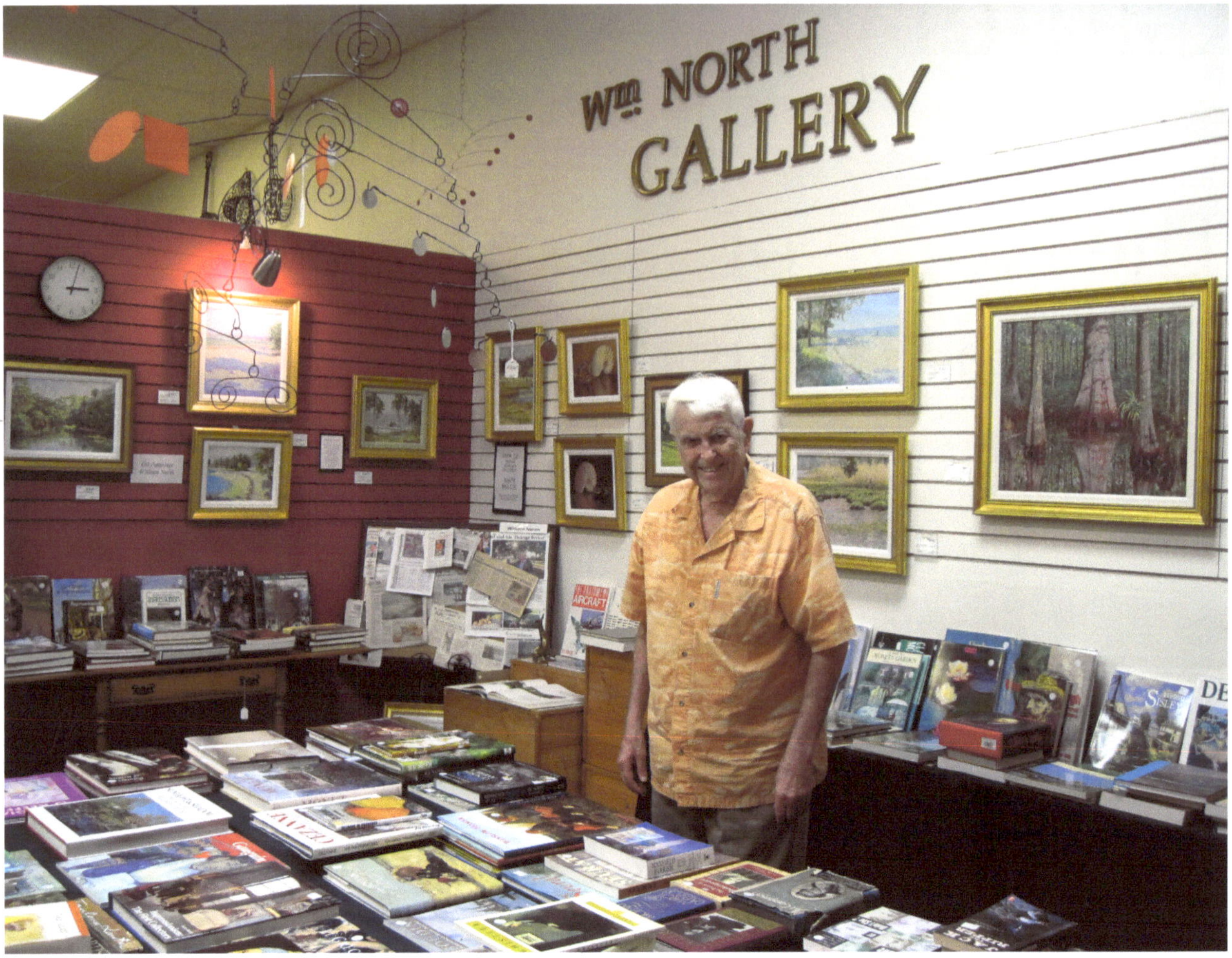

Standing in the gallery, my artist's home away from home.

Wm North

Wm. North Gallery
Inside Gannon's Antiques & Art Mall
16521 S. Tamiami Trail, Fort Myers, FL 33908
(239) 489-2211

Wm. North Gallery

The Home of Florida Citrus Art

Having my own gallery to showcase my citrus art and other paintings was high on my wish list for several years. I dreamed of a home for my new work, particularly my still lifes of Florida oranges and grove scenes.

The dream became a reality in 2004 when an opportunity came along that I eagerly embraced. The Gannons, Ursula, Richard, and son Rick, who had been in the antiques business for many years, were constructing a large twenty thousand square foot Antiques & Art Mall on the corner of US Route 41 and Island Park Road in South Fort Myers. They planned to rent some space to outside vendors and use the rest for their own antiques. I read about their project in the local paper and watched the construction progress as I drove past on my way to and from home.

One day I stopped by, to see what was happening, and met the owner. We discussed the arrangements, and I made a quick decision to open a gallery there. A contract was signed on the spot, and I selected my space (the best, I thought). My art had found a home.

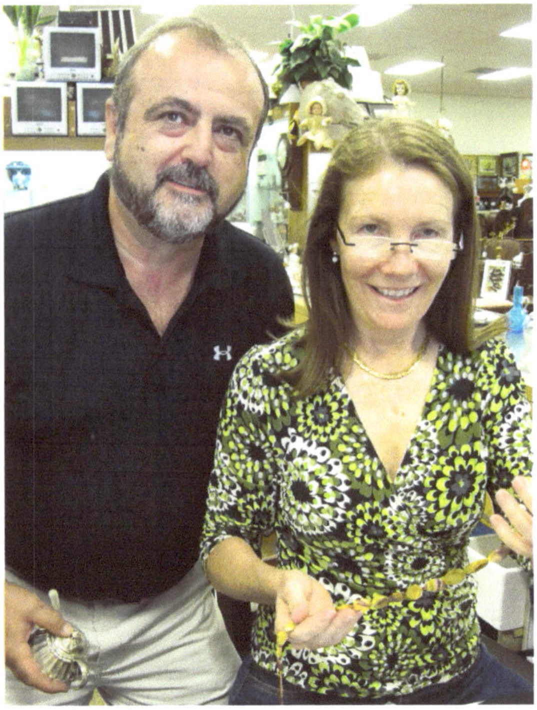

Richard and Ursula Gannon

I asked Carl Schwartz, a painter and faculty member at Florida Gulf Coast University, to join me in the venture. He said "yes" and then suggested we add Bruce Gora, a popular architect, photographer, and active member of the local art community. We named our space the *Florida Fine Art Gallery* and were in business, showing only our own work, except for a few guest artists for one-month shows. In time, Carl and Bruce left to pursue other interests, and the gallery became my baby. I renamed it the *Wm. North Gallery*.

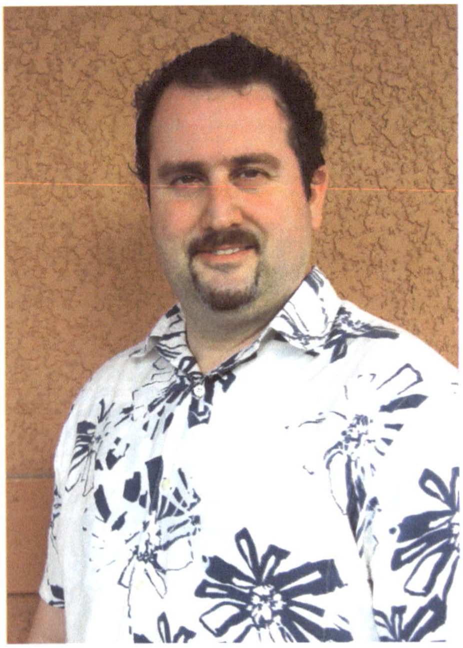

Rick Gannon

To add variety, I invited Ludmila Evans, a world-class artist in clay, to join me and show her pottery and jewelry. Friend and fellow artist, Ralph Bigletti, a long-time metal sculptor, then accepted my invitation to hang his mobiles from the ceiling.

Eventually, Ludmila's success required her to take a larger booth in the mall. Using her vacated space, I added an art bookstore that

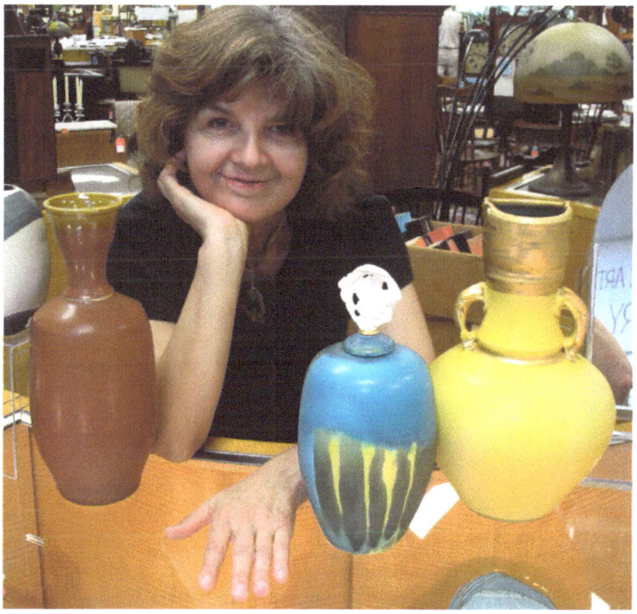

Ludmila Evans

features rare and out-of-print editions. Thanks to the Gannons and their always helpful, congenial staff, the gallery has thrived. In 2011, six successful years later, the *Wm. North Gallery* showcases my paintings, prints, and the art bookstore, plus Ralph Bigletti's mobiles.

Being a gallery keeper has been satisfying from the start, and I am always happy when one of my pieces of art finds a new home. I soon realized that my business was not just a business. I found that standing in my gallery and having far-ranging conversations about art and artists with people I met added a new and stimulating quality to my business experience. These folks, both customers and casual passers-by, came from near and far: some from a vibrant local population, others from a constant tourist flow, and some from among the flocks of snowbirds who appear each year. All added spice to the conversation and flavor to the visit.

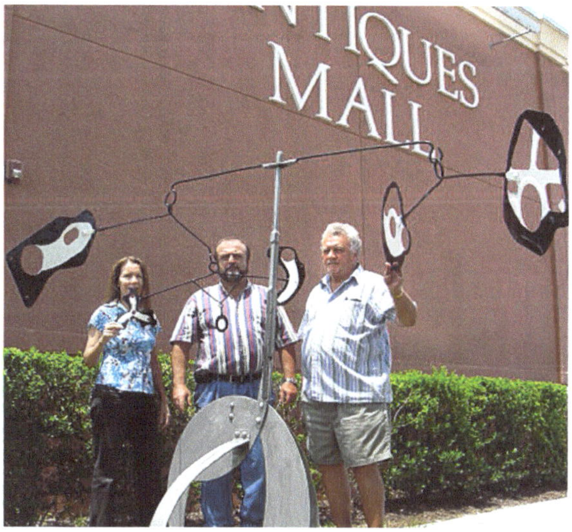

Ursula & Richard Gannon with Ralph Bigletti

It is these people, many of whom have become friends, who keep me active and working at the easel.

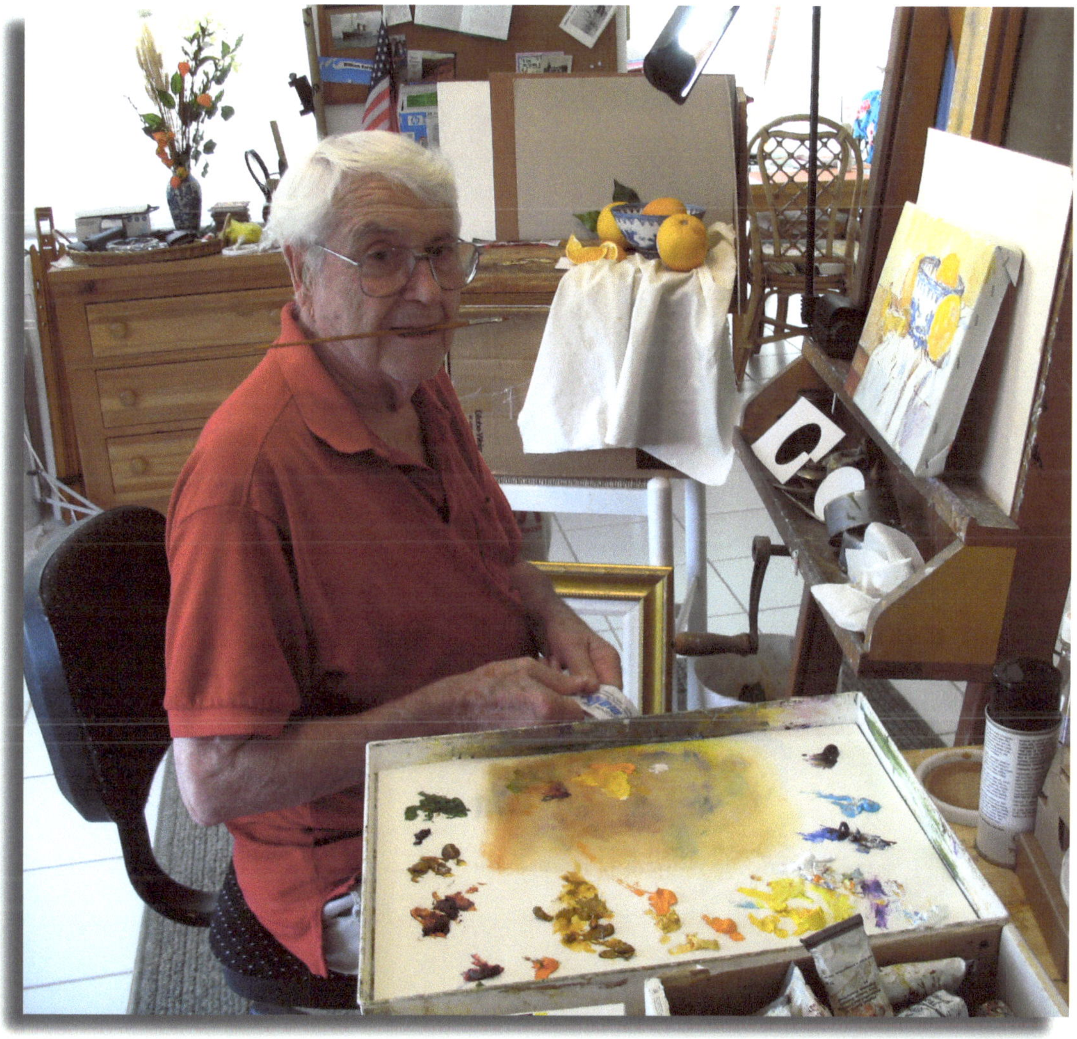

www.ingramcontent.com/pod-product-compliance
Lightning Source LLC
LaVergne TN
LVHW070953210926
848482LV00002B/208